I0619371

The Red Baron vs. Mothman

Published by John J. Rust at Kindle Direct Publishing

Copyright © 2023 John J. Rust

Cover design by C.J. Siegfried

ONE

The pain rocked his skull like blows from a hammer.

Manfred von Richthofen gritted his teeth. He couldn't afford a headache now, not in the middle of a dogfight.

He fought the urge to close his eyes. It felt like his head was about to split apart. But shutting his eyes could mean losing sight of the British bomber in front of him.

Or of another enemy looking to shoot him down.

Grunting against the pain, he raised the bullet-shaped nose of his Albatros D.V. Manfred needed to get above the other plane before commencing his attack. The DH.4 could fly higher than his fighter, but had a much slower rate of climb. He also had to be aware of its rear-mounted machine gun.

Manfred's plane rose. He leaned right, trying to keep the bomber in view.

It ducked into a cloud.

He snorted and leveled out. What direction would the British pilot emerge?

Manfred took the opportunity to glance around him. Alois Heldmann's Pfalz D.III stayed behind him, guarding his tail. Teamwork was one of the reasons his pilots had been so successful.

All around him the brightly colored planes of *Jagdgeschwader 1* twisted and turned as they went after the DH.4s. Manfred glimpsed one of the large biplanes plummeting toward the ground, trailing smoke and flame. He nodded. One less British plane to bomb the German defenses in Flanders.

Manfred looked back to the cloud. The DH.4 barreled out the left side of the white blob. He pulled back the controls. The Albatros roared higher into the sky, ready to dive on the enemy.

Just be mindful of the rear gun. He grimaced, fiery bolts digging into the scar that ran across his skull. A bullet – or piece of shrapnel – had nearly blown out his brains earlier in the month. He had no desire to get shot in the head again . . . or anywhere else, for that matter.

The DH.4 turned south. The Albatros's engine buzzed with the ferocity of a million hornets as Manfred pushed the plane into a dive. Bitter cold wind blasted through the open cockpit and over his head and shoulders. He aimed the nose at a point in the sky where he anticipated the British bomber to be in a few seconds. He leaned forward, staring through the propeller. His thumbs hovered over the buttons for the two forward-mounted MG 8s. Kill number fifty-eight was within his grasp.

A dark shape zipped between him and his target.

Manfred jerked and snapped his head left. He lost sight of . . . the plane? Could it have been a plane? It seemed too small.

But what else could it have been? It certainly did not belong to any of his *Jastas* – fighter squadrons. It had to be British.

Manfred swung his Albatros in the direction of the mysterious airplane. The tactic he preached to his "pups." Never run away from the enemy. Always turn toward them to engage.

He gazed across the sky, but found no sign of the newcomer. He glanced back at the DH.4. The bomber grew smaller as it continued its retreat. Manfred frowned. He'd wanted to shoot down that plane and give some relief to the soldiers below. But he couldn't ignore the aircraft that suddenly appeared. Something that small had to be a fighter. Paying it no mind could be the death of him.

Manfred searched among the blue sky and white clouds. The small airplane was nowhere to be found. He checked behind him. Heldmann remained with him. Whether he had also seen the other plane Manfred had no idea.

A line of black stood out in the sky. Smoke. Another DH.4 had fallen victim to one of his pilots. Three more bombers flew south, back to Allied lines.

Manfred was tempted to pursue them, wanted to have another kill to his credit. But chasing the enemy into their territory meant running the risk of encountering British, French, or Canadian fighters. His men had accomplished their mission. British bombs would not fall on the German lines. It was time to return to base.

The colorful planes formed up behind Manfred's red one. He checked his compass, then landmarks below as he led *Jagdgeschwader 1* west. Twisting left to right in the cramped cockpit, he quickly counted the planes around him. It didn't appear the British claimed any of his pilots.

No. He searched the formation again. One plane was missing, the red and yellow Albatros DIII flown by *Leutnant* Förster.

His shoulders sagged. Another good man lost. Too many good men had been lost over the last several months. Over the last three years.

He could have crashed and survived. If that were the case, Förster was likely badly injured. Either he would be found by a German patrol and taken to a hospital, or found by a British or French patrol and made a prisoner of war.

Or he would succumb to his wounds alone in some God-forsaken crater-filled piece of earth.

Manfred closed his eyes, praying for the best possible outcome for Förster. Deep down, he had a feeling the poor man had become one more number among the hundreds of thousands of Germans already lost in this war.

He soon came upon a large field dotted with tents. Manfred descended. The green grass swelled before him until the wheels bounced upon the ground. The Albatros shuddered. Manfred bared his teeth as invisible fists pounded on his head. He sucked down a couple of deep breaths, fighting through the pain as he rolled his plane toward a large canvass shelter. He waited until the propeller stopped spinning before hauling himself out of the cockpit and jumping to the ground.

His legs quivered. So did his stomach, to the point Manfred had to clench his teeth to keep from throwing up. He pressed a palm against the side of his plane to steady himself.

Why is this happening to me? Ever since he returned to duty barely a week ago, he felt on the verge of collapsing or vomiting, or both, whenever he returned from a patrol. It even happened a couple of days ago when Anthony Fokker himself had come here to have Manfred personally test his new triplane. Thank the Lord he had not gotten sick or crumpled to the ground as the aircraft designer had brought a camera crew to film the "momentous occasion."

"Are you all right, *Herr Rittmeister?*"

Manfred looked up at the boyish, bespectacled mechanic who had addressed him by his old cavalry rank.

"Ja. Ja. No need to worry, *Unteroffizier."*

Unteroffizier – Corporal – Ewers raised an eyebrow, but said, "Very well, sir."

Ewers and a few other mechanics wheeled the Albatros into the canvass shelter. A harsh breath escaped Manfred's nostrils. That was not good. The enlisted and non-commissioned officers liked to gossip. What if Ewers told his friends *Der Rote Kampfflieger* – "The Red Fight Pilot" – looked unsteady when he got out of his plane?

Whatever ailments he had, Manfred had to fight through them. A commanding officer could not exhibit weakness, lest the men under him lose confidence.

He sensed something big approaching. Manfred turned to see a great dane trotting over to him.

"Moritz." He scratched the dog's head. The animal sat beside him as more planes from *Jagdgeschwader 1* landed. Manfred waited for one in particular, a sandy-colored Albatros DIII. The airplane flown by *Leutnant* Gude, Förster's wingman.

He waited for the propeller to stop before walking to the pilot, Mortiz following.

"Gude," he called out as the mechanics pushed the DIII toward a canvas shelter.

"Yes, sir?"

"What happened to *Leutnant* Förster?"

A grimace formed on Gude's round face. "I'm sorry, *Herr Rittmeister.* I don't know. We were pursuing one of the British bombers when I took machine gun rounds to the wing. I turned left to try and throw off their aim. When I swung back around, Förster was gone."

"Did you search for him?"

"Of course. The best I could. But . . . I could not find him." His lips tightened, a dour expression falling over his face.

"So you saw no sign of his plane being shot down?" asked Manfred.

"No, sir. He simply vanished."

Manfred stared off to the side, thinking. At least trying to think through the pressure threatening to crush his brain. He drew a long breath, fighting to push aside the pain.

"He could have become disoriented," said Manfred. "Gotten separated from the rest of us."

"So he might be on his way back here." Gude's eyes lit up at that hope.

"Possibly." Manfred thanked him and went off to question the other pilots. Förster could have trailed far behind the main body of the *Jagdgeschwader.* Another pilot might have seen him where he hadn't.

He asked five men. None had seen Förster following the battle with the bombers. He searched for another pilot to ask when Heldmann hurried over to him.

"Herr Rittmeister. I think that is Förster." He pointed to the sky.

Manfred looked. It was indeed a red and yellow DIII. His stomach uncoiled. Even the throbbing headache lessened. His pup had indeed survived the battle.

His relief evaporated when he observed Förster's approach. Heldmann voiced his concern. "He's coming in too fast and steep."

Manfred tensed, watching the DIII barrel toward the ground, the wings jerking left and right. His fear spiked. Had Förster lived through the battle only to die in a crash?

The worst of it, Manfred could only watch helplessly and pray the *leutnant* came through this alive.

The plane barely slowed, didn't even try to lift its nose. Manfred jerked when the DIII slammed into the ground. Its left landing gear snapped off. The left wing tilted and struck the earth, shattering. The plane spun about ninety degrees, its deformed nose planted in the grass.

Manfred and Heldmann ran toward the wrecked aircraft, Moritz sprinting after them.

Please be alive. He held his breath as he neared the cockpit. Förster was slumped over.

"Förster!" he called out. "Förster!"

The pilot's shoulders slowly rose and fell.

Manfred let out a relieved breath. Thank goodness. Still alive. Though probably injured.

Could he have been injured during the fight? Perhaps that's why he suffered a crash landing.

"Förster? Are you wounded?" he asked as Gude rushed over. Several other pilots ran toward the damaged DIII.

"Förster?" Manfred said his name when he did not answer.

The pilot took another slow breath, then leaned back. Blood ran down his cheek. He didn't appear seriously injured.

But his face. It was as white as a sheet. And his eyes. Wide, staring ahead at . . . nothing.

"Förster? What's wrong?"

The pilot did not acknowledge him.

"Förster?" Manfred reached out and touched his shoulder.

The young man screamed, so loud, so horrific, it made Manfred jump back.

Another prolonged wail threatened to tear apart Förster's throat. Manfred reached in and grabbed the pilot's flying jacket.

"Get a hold of yourself! What's wrong with you?"

Förster looked at him with unblinking eyes. His jaw quivered, then he yelled out the same words, over and over.

"Red eyes! Red eyes!"

TWO

Manfred stood outside the hospital tent waiting for word on Förster's condition. Several other pilots had also congregated around the entrance. While he appreciated their concern for their injured friend, he did not want them to create any distractions for the medical staff and sent them away with the promise he would let them know how the pilot was doing.

He gazed across the makeshift airfield, barely registering the tents, the sandbag walls, and machine gun emplacements. All he could think of was Förster's terrified screams, his ramblings about "Red eyes!" What had he meant by that? The flier had been in much more intense battles than the one today. What could have frightened him to such an extent?

The tent flap opened, which made Moritz, lying nearby, lift his head. Out stepped a tall man with a long, craggy face and steel gray hair.

"How is *Leutnant* Förster, Doctor?"

Dieter Kretschmer, *Jagdgeschwader 1's* staff surgeon, nodded. "He is lucky. I've seen other pilots in much worse condition from such accidents. He suffered injuries to his knee and shoulder, as well as a laceration to his face. Again, not too serious, but he will require time to convalesce."

Manfred stared at the doctor, waiting for him to address the one aspect of Förster's condition that concerned him the most. But Kretschmer remained silent.

"You told me about what ails *Leutnant* Förster's body, but what of his mind? You saw him when we carried him into the tent. The man was in hysterics."

"I did give him chloroform to settle him down."

"Perhaps he is settled now, but why did he react in such a way before? What could have scared an experienced fighter pilot like him so much?"

Kretschmer frowned. The doctor checked over his left shoulder, then his right, as though worried an Entente spy might be eavesdropping.

"May we talk in private, *Herr Freiherr?*" He used Manfred's Prussian title of Free Lord.

That made him cock an eyebrow. Just how serious was Förster's ailment?

"*Ja.*" He nodded. "Come to my tent."

They trekked across the field, Moritz trotting beside his master. Kretschmer lit a cigarette before they entered the tent. It was a simple setup with a cot, a trunk, a desk, and two stools made out of wooden crates. At the far end was a stand with dozens of silver cups, specially made to commemorate each of Manfred's fifty-seven victories.

He seated himself behind his desk, the great dane laying down nearby. "So, Doctor. What is it about Förster that you feel must not be overheard?"

Kretschmer stared at the cigarette between his fingers, his shoulders rising in a slow breath. He pressed his lips together, looking hesitant to speak.

After a few seconds, he finally looked up at Manfred. "Tell me. Have you heard of the term 'shell shock'?"

Manfred's brow wrinkled. "No."

"It's something the British came up with. A more clinical term would be neurasthenia."

Manfred gave a derisive chuckle. "You doctors have a talent for coming up with complicated words. So what exactly is this . . . shell shock, as the British call it?"

"They have written about it in some of their medical journals. It describes soldiers who can no longer handle the strain of battle. Symptoms include nightmares, facial tics. Some reports indicate soldiers have lost the ability to speak or become paralyzed, even though they suffered no actual injuries that would render them that way. To put it simply, they undergo a nervous collapse."

Manfred grunted. "Perhaps they are weak men. Or they have no honor and pretend they are hysterical so they do not have to fight for their country."

"Would you use either of those characteristics to describe *Leutnant* Förster?"

"Of course not." Manfred snapped his hand in a dismissive gesture. "Why, the man had five kills during Bloody April. He terrorized the Entente, took pride in his fighting prowess. No, Doctor. *Leutnant* Förster is not someone who would have a nervous collapse, as you call it."

"Then I do not know how else to explain it. Nothing about his injuries would account for his behavior when he landed." Kretschmer took a drag on his cigarette, then stared at the silver cups for a few moments. "Maybe it is an accumulation of all the battles. Maybe there is only so much fighting a soldier can experience before it overwhelms his mind, his very soul."

Manfred snorted. "A soldier's role is to fight. He knows this."

"True, but things are so different now."

"No they are not. War has been a constant throughout human history."

Kretschmer exhaled and shifted in his seat. "If you will, *Herr Freiherr,* consider this. Go back to August 1914 when this war began. What was the general attitude? Our army would march across the fields of Europe, stand in neat lines before the enemy with rifles at the ready, while the cavalry made a heroic charge with swords and pistols in hand. Somewhere behind these formations, colonels and generals would sit up straight on their horses and command their troops. And everyone would be home by Christmas."

He lowered his gaze. "A couple of weeks from now, we mark the third anniversary of the start of this war. What have we seen in those three years? Millions of soldiers stuck in muddy, stinking, disease-ridden trenches for months, subjected to constant barrages of bullets, artillery, and gas. Friends dying all around them, while they spend every minute fearful a bullet may rip through their chest, or a shell may blow them to pieces, or chlorine gas will dissolve their lungs. Even worse, there is no end in sight to this misery. How long can any man cope with all that before he breaks down?"

Manfred folded his hands and stared at the staff surgeon, contemplating his words. Yes, warfare was brutal and terrifying. But a soldier had to be strong and deal with to bring victory to his country.

Though he had to admit Kretschmer had a point. Three years was a long time to be in constant combat. Wars of the past would normally see a battle last a few days at most, followed by a lengthy period for both sides to regroup before the next engagement. There were no battles like Verdun, which lasted nearly all of 1916 – with the French victorious, curse them. No living in trenches for months at a time. Certainly no gas. And the number of dead! Germany had lost four times the number of men in four months of fighting in the Somme last year than it had during the entire Franco-Prussian War more than forty years prior.

I guess that could be overwhelming to a lot of men. Still, a soldier's duty . . .

Kretschmer grunted. "Of course, the colonels and generals see none of this. Instead of being on the battlefield with their men, they commandeer mansions and run the war a hundred miles from the nearest bullet."

Manfred steepled his fingers. "What you speak of is what the infantry goes through. We do our fighting in the skies, not on the ground."

"Which is no less risky," countered Kretschmer. "How many pilots have you lost over the last year?"

"Too many." Manfred's shoulders sagged.

"And you do not think that weighs on the minds of your men? Maybe to the point where --?

"Nein!" Manfred thumped a fist on his table. "I only pick the best pilots for my squadrons. Men who have fought in numerous battles and have shot down many planes."

"You know as well as I do even the best pilots can be shot down."

The corner of Manfred's mouth twitched. He wanted to say something to contradict the staff surgeon, but couldn't. All he had to do was think of his mentor, Oswald Boelcke. The man had forty victories to his credit and came up with *Dicta Boelcke*, the manual of successful fighter tactics. Manfred lived by those teachings and passed them on to his pups.

But all his brilliance and skill could not save Boelcke. During a battle last October, his plane collided with another, causing him to go down near an artillery battery. Everyone said he should have survived the crash. At least, he would have if he'd been wearing his crash helmet and safety belt.

How could someone that experienced make such an error in judgment?

Manfred's jaw stiffened. Could Boelcke have suffered from some sort of nervous collapse prior to his last fight? He recalled his mentor not acting like himself in the days before his death.

No! A man like Oswald Boelcke could never be that weak.

But how else could he explain the man not taking the simplest of precautions before taking off?

Manfred tapped a finger on his table, thinking about Förster's dilemma. It was much easier to do so with his headache nearly gone.

"Let's say, for argument's sake, I accept your explanation of shell shock. How can we cure *Leutnant* Förster and get him back in the sky?"

"I wish I had an answer, *Herr Freiherr*. I have heard of some doctors who use . . ." Kretschmer's head bobbed from side-to-side, as though searching for the right words. "Rather strong methods to treat afflicted soldiers."

"Do they work?"

"From the little I have heard, that is up for debate." Kretschmer took a puff from his cigarette. "Förster still needs to recover from his injuries. I'll see he does that back in Germany. Perhaps being far removed from the frontline will improve his state of mind."

"Let us hope so, Doctor," replied Manfred. "The latest intelligence I have received indicates the British are preparing for another major offensive in Flanders. I need every pilot available for the coming battle. Keep me posted on *Leutnant* Förster's progress."

"I will, *Herr Freiherr*."

"Good. You may return to the hospital tent."

Kretschmer stood. He took a step toward the tent flap when Manfred said, "And Doctor. You were right to share this with me in private. Let us keep this talk of shell shock between us. It might worry the men that they could one day suddenly start screaming hysterically like *Leutnant* Förster."

"As you wish, *Herr Freiherr*."

The Doctor took a couple of steps before Manfred halted him again. "One more thing. Förster kept screaming about red eyes. Those were the only two words he said. Why red eyes?"

Kretschmer glanced to the side in thought. "The only explanation I have is that he may have been hallucinating."

"Is that another symptom of shell shock?"

"Perhaps. In all honesty, this is a new malady. We doctors are still learning how it afflicts soldiers. But given what I know about it so far, it would not surprise me."

Kretschmer reached the flap, then paused. Maybe he thought Manfred was going to ask him another question.

He didn't, and the staff surgeon exited.

Manfred sat at his desk, staring quietly at the other side of the tent. He thought of Boelcke, of Förster. He couldn't recall the former being downtrodden anytime over the past couple of days. Quite the contrary, Förster seemed in high spirits. So what would cause him to suddenly have such a breakdown?

And possibly hallucinate.

Manfred thought back to today's battle with the British bombers, and that strange, small aircraft he glimpsed. He closed his eyes, trying to recall its details. It was dark, like a thick shadow. The wings. They seemed . . . moth-shaped.

The more he thought about it, the more it seemed it hadn't been an airplane, but a . . .

No. That is not possible.

But why couldn't he find it after his brief encounter? He should have been able to spot it. Unless . . .

Oh God. Please no.

What if he had begun hallucinating like poor Förster?

THREE

Another new pilot.

Manfred softly sighed as the lanky young man walked into his tent. How many meetings had he had with new pilots over the past few months? Too many. The attrition rate among *Luftstreitkräfte* fliers was atrocious. He fought to keep a grimace off his face, wondering how many more replacements he would have to greet in the coming weeks.

The pilot stopped near Manfred's desk, ramrod straight, his eyes wide with awe. His lips parted, but it took a moment for him to find his voice.

"He-Herr *Rittmeister. Leutnant* Fritz Erhardt reporting as ordered."

"*Leutnant.*" Manfred nodded. "Welcome to *Jagdgeschwader Ein.* Have a seat."

"Thank you, and may I say, it is an honor to be part of your squadron."

Erhardt sat on the stool, eyes still wide, a smile creasing his lips. It was not the first time someone reacted this way in Manfred's presence. He could thank the newspapers and magazines in Germany for that. Their stories had not just made him famous, they elevated him into an almost legendary hero on par with those from Greek or Teutonic mythology.

Luckily, this excessive admiration by new pilots faded after a day or two when they realized he was an actual man, one who demanded the best from them. He certainly stamped it out before the pilot's first patrol. He could not afford someone in the air enamored with the idea of flying with "The Red Fighter Pilot."

"It is my hope that you will bring honor to this unit," said Manfred. "That is why I requested you."

"Of course, *Herr Rittmeister.* I will not let you down."

"You mean you will not let *us* down. As in, the entire wing."

Erhardt gave a nervous twitch of his cheek. "Um . . . Yes. I meant I will not let the wing down."

"I should hope not." Manfred leaned back in his chair. "Especially given your combat record. Twelve kills since your first squadron assignment in February. Quite impressive."

"Thank you, sir."

"And three of them balloons. Again, quite impressive."

"Thank you, sir." Erhardt beamed.

The young man had every right to be joyous over that, thought Manfred. Shooting down a balloon may not sound as romantic as gunning down a fighter plane. But these balloons were not the kind a child might enjoy. These carried observers who spotted for artillery and reported troop movements. They were also well-protected by anti-aircraft guns. Destroying just one took great skill and bravery, and *Leutnant* Erhardt had downed three.

But also paid a price.

"You were wounded in the leg when attacking one of those balloons. How is it now?"

"Fully recovered." Erhardt slapped the side of his right leg, tacking on a smile. "God has blessed Germany with some of the most gifted doctors in the world."

"Mm-hmm." Manfred clasped his hands together and rested them on his desk, still eyeing the other pilot. "You were also injured after an engagement in May, correct?"

"Yes, *Herr Rittmeister*. A Canadian flier snuck up behind me and practically shredded my right wing. I still managed to return to base, but crashed my Albatros. I fractured my ankle and dislocated my shoulder, but it could have been much worse. God was watching out for me that day."

Erhardt reached into his tunic and produced a small Bible. He held it up with a satisfied smile. "I always pray before taking to the air."

"I see." Manfred nodded. "Very wise, especially in a war like this. Though perhaps God would like you to do more than pray to him."

"Sir?" Erhardt's face wrinkled in puzzlement.

"Perhaps he wanted to make your crash a teachable moment. Your former commanders praised you for your aggressiveness, but what they seemed to have ignored was your recklessness in combat."

Erhardt stared at him, mouth agape. "Sir? I . . . I do not understand."

Manfred locked eyes with the other pilot. "How long were you pursuing your target before that Canadian fighter shot you?"

"Um . . ." Erhardt looked to the side in thought. "A couple of minutes, I would guess. That pilot was twisting and turning like mad. I had a hard time getting him in my sights."

"And what else was going on around you while you chased after that plane?"

"I'm not sure. I was trying to shoot him down."

"And that is why you were almost shot down." He jabbed a finger at Erhardt for emphasis. "You fixated on your target, ignoring everything else around you. That is how the other Canadian managed to slip behind you."

Erhardt opened his mouth, but stayed silent, as though unable to form a response.

"Perhaps God spared you so you learn from your mistake," Manfred continued. "And you will learn in this wing. Getting caught up in a drawn-out dogfight is a sure way to wind up dead. Here we go for the quick kill. The kill where the enemy pilot does not even know we are coming after him. We also fight as a cohesive unit. We protect each other in the air, make sure no one is sneaking up on you as you go in for the kill. You cannot guard your fellow pilots if you go off in pursuit of a single plane. Do you understand?"

"Y-Yes, *Herr Rittmeister.*" Erhardt lowered his gaze, looking like a scolded child.

That seems to have rid him of the awe of being in the same tent as Der Rote Kampfflieger.

"You are already an excellent pilot, *Leutnant* Erhardt. Follow my teachings and those of your squadron leaders, and you will be an even better pilot, with better odds of surviving this war."

Erhardt lifted his chin, perked up. "Yes, *Herr Rittmeister.*"

Manfred nodded. While the man had been stung by the criticism, he seemed willing to listen to any advice to improve his flying.

He started to go through the *Dicta Boelcke*. Not only would it benefit Erhardt, but it benefitted Manfred as well. It helped get his mind off the worry that he might be hallucinating due to shell shock like poor Förster. But as he had not seen any more small, moth-winged aircraft since that day, he figured he was all right.

He hoped.

Manfred pushed aside the thought and explained to Erhardt how he should only fire when in close range of the target. That's when a buzzing sound filtered into the tent.

"What's that?" Erhardt scrunched his face and turned to the tent flap.

Manfred followed his gaze. "I believe that is our other new addition to *Jagdgeschwader Ein.* Come."

The two exited the tent. Moritz, lying outside the tent, rose and followed his master. Manfred's eyes locked on a motorcycle with a sidecar rolling toward him. At the handlebars was a lean man with a narrow, boyish face. He couldn't help but grin as the rider neared.

The motorcycle jerked to a stop a couple of meters from him.

"Manfred!" The younger man hopped off and spread his arms.

"Decorum, *Leutnant.*"

"Oh." The rider grimaced. "Yes, you're right." He cleared his throat and straightened. "*Leutnant* Werner Voss reporting as ordered to join the illustrious ranks of *Jagdgeschwader Ein.*"

Were it anyone else, Manfred would have reprimanded him for such an inappropriate response. But he always had a soft spot for Voss and his eccentricities.

"And I am glad to have you in our ranks." Manfred stuck out his hand. "It is good to see you again."

"As it is you, Man . . . er, *Herr Rittmeister.*" Voss shook his hand. "And I brought something to celebrate our reunion."

He reached into the bag in the sidecar and produced a bottle of red wine, which he held up with a big smile.

"Where did you find that?" asked Manfred.

"I ran into a patrol not far from here. A sergeant was carrying this on him. I had to give up two packs of cigarettes, all my coffee, and a chocolate bar for this bottle. Greedy bastard."

"It will make a fine addition to dinner tonight."

"That it will." Voss turned to Erhardt. "And who is this?"
Manfred introduced them.

"You must be good if *Der Rote Kampfflieger* picked you for this unit," said Voss.

"Twelve kills to my credit." Erhardt lifted his chin with pride.

"Not bad. I have thirty-four myself. We'll see if you can catch up." Voss tacked on a grin.

"Come," said Manfred. "I'll show you to your tent."

Voss grabbed his bag and followed, gazing around the airfield. "I hope you have my quarters clearly marked. It could be hard to find among all these tents. Heh! It reminds me of a big circus."

"You are not the first one to make that observation," Manfred said over his shoulder. "Luckily, you should not have a problem finding your quarters. Each *Jasta* commander has his own tent."

"*Jasta* commander." Voss snorted and shook his head. "How could you do that to me, your good friend? Foist such a responsibility on my shoulders."

"You act as though this is a surprise. You have previously commanded two squadrons."

"Temporarily." Voss scrunched his face. "And I hated every minute of it. Do you know how much paperwork you have to do as a squadron commander?"

Manfred cranked an eyebrow. "I have some idea, yes."

"Bah." Voss shook his head. "I would rather concentrate on shooting down the enemy and not writing more than a newspaperman."

Manfred stopped and turned to face him. The concerns he had about handing command of *Jasta* 10 to Voss resurfaced. Though a naturally gifted flier, his friend had little use for discipline and protocol, to the point he tried to get the commanding officer in their former squadron removed on the grounds the man lacked aggressiveness. Then there was his age. Though experienced, he was just twenty years old. How would the older pilots react to taking orders from someone barely in adulthood?

In the end, Voss's positive traits outweighed his negative ones. "I would not have made you a squadron commander if I did not think you were qualified. You have proven yourself numerous times in combat, and right now I need men like you. The British are preparing for a major offensive. They are sending more and more bombers and reconnaissance planes over Flanders, along with fighter screens to protect them. We need to control the skies, to delay or even stop the British offensive, and perhaps launch one of our own. With the Americans declaring war on us, we need to defeat the Entente as quickly as possible."

Voss scoffed. "What do we need to fear from the Americans? Their army is small and they haven't done anything of note since defeating Spain twenty years ago."

"Their army will not be small for long," said Manfred. "America's population is greater than ours. By this time next year, they could have millions of men mobilized against us."

Voss winced. "Yes, that would not be good. So I guess the fate of the German Empire is in our hands."

"We certainly have a critical role to play."

"Uh-huh." Voss slapped Erhardt's shoulder, making him jerk in surprise. "Then you can count on myself and *Leutnant* Erhardt here to do our part to preserve our nation and the rule of the Kaiser."

"I expect no less."

They arrived at Voss's tent. He poked his head through the flap, waited a few seconds, then pulled out. "I think this will do."

"I'm happy it meets with your approval."

Voss grinned and clutched Manfred's left arm. "With so much at stake, as you say, I think there is one prediction I can make with absolute certainty."

"And what is that?"

"We are bound to have some memorable adventures."

FOUR

The wine provided by Voss made dinner a more palatable affair. The sweet flavor was a stark contrast to the dryness of the food on Manfred's plate. He did not attempt to keep the frown off his face as he stared at tonight's offerings. A piece of sausage, a few carrots, and bread. Meager compared to the meals he had eaten just a year ago. This even as most of Germany's food stocks went to the military. Another distressing sign as to what this war was costing his country.

"Herr Adam," he heard Heldmann call from a few seats away. "You have barely sipped your wine. Are you going to nurse it through tomorrow?"

"It is called savoring." Hans Adam, sitting across from Heldmann, sniffed his cup, eyes closed in apparent delight. At thirty-one, he was one of the oldest men in the unit, seven years Manfred's senior.

Adam glanced at his plate and grunted. "It is the only thing at this table worth savoring."

"Ha!" barked Heldmann. "On that, we agree. It is bad enough the generals cut our rations, but what we do get tastes horrible."

"If you think this is bad, imagine what it is like for civilians." Manfred stabbed a carrot with his fork and popped it in his mouth. So bland.

"Yes, I've heard the stories. Empty shelves at stores, soup kitchens." Heldmann shook his head. "Damn the British and their blockade. Their ships sail about the North Sea as they please, and where is our navy? Sitting on their asses in port."

"Seems a waste to me," said Adam. "Building all those battleships and cruisers and not even using them."

Manfred gave a slight grunt in agreement. He knew the British Navy was much larger than theirs. Still, he felt it pointless to have invested so much time and money into such powerful weapons and barely use them. It appeared it would be up to the Army and the Air Service to win this war, and win it quickly.

He looked down at his plate, wondering how small the portions would be if the fighting dragged on for another year.

Manfred took a swallow of his wine, losing himself in the red liquid's sweetness. Forgetting about food shortages and shadowy, moth-winged hallucinations, if only for a second.

When dinner ended, Manfred strolled around the base before returning to his tent. His stomach wasn't full, but neither was he hungry. Again, he thought of the populace in Germany. How many were going to bed tonight starving? Probably not his family, being of Prussian nobility. Though his mother had told him during his convalescence that their cupboards were not as full as they had been just two years ago.

Would those cupboards be empty next year?

That worrying thought made him remove a piece of paper from his desk drawer. By candlelight, he composed a letter to his parents. From outside, the mournful tone of a lone violin floated across the air. *Leutnant* Gude, with his nightly one-man concert. Tonight's selection, "Siegfried's Rhine Journey" by Wagner.

Manfred's pen hovered above the paper as he stopped writing. He stared at the partially open tent flap, absorbed in Gude's playing, especially when he got to the crescendo. Like the wine at dinner, another means of escape from the war, the lack of food, the hallucinations.

The man is incredibly talented. Not surprising as Gude had been a member of the Leipzig Orchestra before the war. Perhaps once the fighting was over, he could see Gude perform with them again.

Provided we both survive this –

Muffled snaps mingled with the energetic strings. Manfred's brow wrinkled. Those snaps merged into a constant, distant crackle, more distinct now that Gude had ceased playing.

There was no mistaking that sound. Gunfire.

Manfred dropped his pen and hurried out of the tent. Several other pilots and mechanics stood outside, staring out at the darkened fields that surrounded their base. The shots remained distant. Perhaps a couple of kilometers away, he guessed.

"Who could be shooting?" asked Gude.

"It might be that patrol I came across earlier today," offered Voss.

Heldmann stepped over to Manfred, head cocked to one side. "But who could they be shooting at?"

More steady cracks echoed from the shadowy horizon. "It could be an enemy raiding party."

"And they're probably not raiding the local farms for fresh milk," Voss chimed in.

"No." Manfred swung around to face his men. "Blow out all candles. I want this base pitch dark. Then get to the armory tent. Von Döring . . ." He looked to the blond, round-faced *Oberleutnant* who had commanded the wing during his absence. "Take *Jastas* Four and Six and guard our planes. If there is a raiding party out there, that will undoubtedly be their objective. The rest of you will come with me to defend the base perimeter."

The men hurried off. Manfred dashed into his tent and blew out the candles on his desk and by his cot. When he ran back outside, the orange hues coming from other tents winked out one after the other. Soon the entire air base was plunged into darkness. Men and tents became silhouettes. The moon was partly covered by clouds, further obscuring any potential illumination.

Booted feet pounded on the grass. Even without any light, Manfred knew the layout of the base by memory. Most of the other pilots likely did as well. They sprinted for the armory tent. He imagined the new men like Voss and Erhardt just followed the mass of blackened human forms.

Someone threw back the tent flap. Gewehrs clacked against their wooden racks as men grabbed them. Unfortunately, there were not enough rifles for all the pilots. Some would have to rely on their pistols if the British attacked.

Gewehr in hand, and five-round stripper clips stuffed in his pockets, Manfred dashed toward the front of the air base with *Jastas* 10 and 11. It was then he noticed the rifle fire had tapered off. He hoped that meant the raiding party had been driven away.

But it could also mean the enemy had defeated their patrol and were making their way here.

Manfred reached the sandbag wall and deployed his men, Gude on his right and Erhardt on his left. Everyone moved efficiently. No hesitation, no panic. As he expected. Most of the men in *Jagdgeschwader 1* had previously served in the infantry or, like him, the cavalry before joining the Air Service.

About a minute passed without any gunfire. Then two. Manfred peered over the sandbags. He saw no movement in the darkness. Neither did he hear anything, at least coming from the field before him. Among his fellow soldiers he caught the sharp intake of a breath, or a brief shuffle of someone adjusting himself, or his own heart.

"Maybe our patrol drove off whoever was out there," Erhardt spoke in a normal tone of voice.

"Quiet," Manfred scolded him with a harsh whisper. "We do not know that for certain." And if an enemy raiding party was out there, he did not want them keying in on the voice of any of his men.

He gazed back and forth along the ground. Any British or French soldiers out there would be crawling, not tromping straight at them. Though with the darkness, he might not see them until they were directly in front of him.

Manfred had no idea how much time passed. It felt as though the minutes dragged. He dared not check his watch and risk missing the slightest movement of something in the shadows.

He took steady breaths, pressing his fingers against the Gewehr's barrel. Sweat beaded on his forehead. His eyes continued to sweep across the darkened field. All was still.

Could Erhardt be right? Could that patrol Voss had encountered earlier beaten off the raiding party? What if the enemy had slaughtered the patrol? Would they disengage, having alerted their target, or press on?

Manfred stifled a grunt. Too many unknowns. All he and the others could do was man their positions and wait. For the rest of the night if need be. He'd be damned if he'd allow these raiders to destroy even one of his planes.

His eyelids drooped, the long day and the lengthy period of inactivity catching up to him. He bit down on his lower lip, trying to use the stinging pain to keep from giving in to his fatigue. Manfred studied the field. Nothing moved. But could his tiredness have caused him to miss something?

And if I am getting tired, so must my men.

Maybe that was the raiding party's plan. Wait for several more hours. Wait until he and everyone else was exhausted and lulled into idleness from staring at a dark, empty field with nothing happening. Then they would –

"Someone is out there."

The harsh whisper jerked Manfred out of his reverie. That hadn't been one of his pilots. The gruff voice belonged to Sergeant Holm, who was in charge of security for the air base.

"Where?" Manfred whispered back.

Holm pointed. Manfred squinted, scrutinizing the darkness. He saw nothing. Maybe the sergeant was imagining it.

No! There. He just caught sight of . . . something. A shadow swayed within the blackness.

Then he couldn't see it. Manfred tensed, holding his breath. Could he be hallucinating?

The shape moved again. Then he heard something. A shaky breath.

Other men must have heard it, too. They brought up their rifles and aimed at the form. Manfred looked down the line. Worry shot through him.

"Heldmann," he whispered as loud as he dared. "Have some men cover other parts of the field. That can't be the only soldier out there."

"Ja." Squatting, Heldmann moved among the pilots, pointing out fields of fire to them.

Manfred nodded, pleased that in spite of being out of the infantry or cavalry for a long time, they still had not completely forgotten how to fight on the ground.

"Everyone hold fire until we identify whoever that is," said Holm.

Manfred did not contradict the sergeant. For all he knew, that could be one of their soldiers out there.

He caught another shaky breath from the darkness, then a moan. The shadow plodded forward, holding something long in its right hand. A rifle.

It had to be one of their soldiers. No enemy soldier would simply tread up to a German air base by himself.

A distraction? Maybe the raiding party hoped all their attention would be on this man and launch their attack.

He gazed around the field. Manfred saw nothing else move. But that did not mean no British soldiers were slithering through the darkness.

"Sergeant," he whispered to Holm. "Prepare a flare. We need to make sure this person is alone."

Holm nodded and withdrew his flare gun. Manfred did likewise, then reminded everyone to hold their fire until they knew who it was approaching the base.

Using his fingers, he counted down from three, then raised his flare gun and fired. Holm followed a split-second later.

Brilliant white light illuminated the field. Manfred saw the figure clearly. He stood about twenty meters away, wearing a gray tunic and trousers and a "coal scuttle" helmet. In his hand was a Gewehr rifle. The soldier stood still and gawked at the flares overhead.

"He is one of ours," said a soldier to Manfred's left. The young man got up.

"Get down, you idiot!" Holm grabbed his shoulder and pushed him back behind the sandbags. "The British could be using him to draw us out and blow out what little brains you have."

Holm leveled his rifle over the wall. "Identify yourself!"

The lone soldier lifted his head, mouth open. To Manfred, he appeared dazed.

"I said identify yourself," Holm demanded.

The other man's shoulders rose in a slow breath. He remained silent.

"Who are you? Speak, or you will be shot!"

The soldier shuddered. His mouth opened and closed. Still, no words came out. Manfred glanced at the sergeant, expecting a shot any moment.

"P-P-Private Schaffner," his voice cracked. "Private Schaffner."

Holm did not lower his rifle. "Private Schaffner, advance. Do not make any sudden moves or you will be shot."

Manfred cranked an eyebrow. The sergeant was being rather paranoid. But he had, by some miracle, survived Verdun. The man had every reason not to take any chances.

Schaffner stumbled forward. When he reached the sandbag wall, he collapsed onto it. A couple of soldiers grabbed him and pulled him behind the cover.

Manfred hurried over to Schaffner. The young man stared into the sky with wide eyes, jaw trembling. Manfred's shoulders knotted. It reminded him of how Förster looked before he went hysterical.

"Private, what were you doing out there?" he asked.

"Pa . . . Patrol."

"Where is the rest of your patrol?"

A shiver racked Schaffner's body. He let out a whimper.

Holm grimaced. "I've seen this plenty of times in the trenches. Soldiers who've seen too much, their mind just shatters."

Manfred said nothing, just thought back to Dr. Kretschmer's talk about shell shock. He looked back down at Schaffner. He still had not answered the question. At least verbally. Judging by his reaction, Manfred guessed the rest of his patrol would not be making their way here.

"Who attacked you?" he asked.

Another whimper crawled out Schaffner's throat.

"Who attacked you?" Manfred spoke in a firmer voice. "The British? The French?"

Schaffner let out a sputtering breath. "R . . . R . . ."

"Who?" Manfred grabbed the soldier's collar. "Tell me."

"R . . . Red eyes."

FIVE

The engine growled louder as Manfred dove on the other plane. He leaned forward, gazing through gunsights. The biplane grew before him. Any second . . .

The other aircraft banked left.

Dammit. The rear observer must have spotted him. It would not be too difficult since the overcast denied Manfred his preferred tactic of diving out of the sun on his opponent.

Flecks of orange flashed from the rear of the British Bristol fighter. Machine gun fire. He snapped his Albatros D.V. into a hard right, gritting his teeth. He'd have to come around to re-engage the Bristol or try to catch another enemy unaware.

Planes buzzed and whipped across the Belgian skies. One spiraled toward the ground trailing smoke. Manfred could not tell if it was British or one of his planes. He glimpsed Adam's Albatros ahead of him chasing a Bristol. To his right, another Bristol jumped Heldmann's aircraft. Gude streaked in and fired. Splinters exploded off the side of the British plane. It tipped over on its left wing and dropped from the sky.

Manfred searched around him. *Where the hell is Erhardt?* He'd assigned the new pilot as his wingman to observe firsthand how he handled himself in battle.

He caught Erhardt's black and orange Albatros DIII slipping behind a Bristol. The young pilot fired. *Too soon,* Manfred scowled.

The rectangular biplane swung left. Erhardt pursued it. Another tight turn by the enemy. Erhardt matched it and fired. It appeared he missed.

It also appeared he developed target fixation, the one thing Manfred always warned his pilots to avoid.

The Bristol climbed. So did Erhardt. The enemy then twisted right, as did Erhardt.

Something flashed to Manfred's left. Another Bristol roared in, making straight for Erhardt.

Manfred angled his Albatros right, aiming for the side of the Bristol. He risked a glance at Erhardt's plane. He still chased after his target, unaware of the threat from behind.

Manfred swung his head left to right, making sure no enemy aircraft came after him.

A bolt of pain tore through his head. He shut his eyes tight and bared his teeth. *Not now. Dammit, not now.*

He struggled to open his eyes, his lids like lead weights. He viewed the scene before him through half-blurred slits. He had closed with the Bristol, coming up on its rear. He fought to ignore the thumping in his skull, to look through his gun sights and –

The Albatros rattled. Manfred's eyes snapped fully open. Three holes had appeared in the upper wing, courtesy of the Bristol's rear gunner.

He hit the fire button, violating one of his tenets. Never fire until your opponent was clearly in your sights.

His rounds missed. The Bristol bore in on Erhardt.

Manfred exhaled loudly. He dipped his wing left . . .

Then banked right. The rear gunner fired, his shots going wide.

The gunsights lined up on the back of the fuselage. Manfred fired. The two machine guns on the cowling chattered. Bits of wood jumped off the Bristol's rear.

The British pilot banked left, trying to get away from Manfred. He pursued, gazing through the gunsights. Off target. He drifted left. Almost . . .

The Bristol climbed. Manfred snorted and pulled back the control stick. The enemy plane charged toward a mass of white clouds. His shoulders tensed. He had to catch him before he disappeared into the cloud bank.

Come on. Come on, he urged his Albatros. It edged closer to the Bristol. Not close enough to allow for a shot.

Manfred glanced past the other plane. It was only seconds from entering the clouds. He gritted his teeth. Now or never.

The twin machine guns rattled. The veins in his neck stuck out, waiting for smoke to spew from the Bristol.

None came. The British fighter raced into the cotton-like clouds and vanished.

"Dammit," hissed Manfred. He kept the Albatros's nose aimed at the cloud bank. A risk, for sure, but he was not going to stay in the clear and risk the Bristol diving on him from who knew which direction.

A blanket of thick white surrounded him. He could only see a few inches in front of him, if that. Manfred swung to the left, cold pinpricks sweeping over the skin under his flight jacket. Would he hit the British plane? Had it already dived to try to attack him?

He plowed through the clouds, counting to five in his head, then pointed the nose down. The white mass vanished around him. Below, planes whirled about in a chaotic and deadly dance. Two columns of smoke snaked toward the ground.

Manfred searched around him for any sign of the Bristol he'd been pursuing. He couldn't find it. Maybe it had broken cover to try and jump him. Or maybe it still hid in the clouds.

He swung the Albatros around, this time skimming the bottom of the cloud bank. Wisps of white passed by him. He gazed through the breaks in the clouds, trying to spot any Bristol he could catch unaware.

There! A lone plane flying away from the whirlwind of fighters. Either fleeing or trying to get into position for an attack. Manfred had no idea and ultimately it did not matter. He had his prey.

Something crossed his vision.

Manfred jerked. So did the Albatros. He sucked in a quick breath and righted the plane. His eyes darted behind his goggles. What the hell was that?

He caught a shadow just below the clouds ahead of him. He gaped at it unblinking. Ink black with moth-like wings.

It cannot be.

He blinked once, twice. The shadow stayed in front of him. Whatever it was, it was real. It had to be. If it wasn't . . .

Manfred exhaled, wrinkling his brow. He stared through the gunsights. Part of the shadow crossed the left ring. Without hesitation he hit the fire button.

The shadow snapped up and disappeared into the clouds.

Manfred lifted his head, swinging it left to right. He saw no sign of . . . whatever it was.

He whipped his plane around, not wanting . . . it to drop behind him.

Another frantic search of the clouds came up with nothing. Was that thing still hiding, waiting to ambush him?

Or what if it was all in my mind? The thought made him shudder.

Planes started breaking off. Probably running low on ammunition or fuel, or both. Manfred dove away from the clouds. He took a couple of looks over his shoulder. No moth-winged shapes chased after him. He did not know whether to be relieved or distressed by that.

Jagdgeschwader 1 formed up for the journey back to base. Manfred took count of his pups. Four planes were missing. His stomach caved in. The British had exacted a heavy toll this day.

He stared straight ahead, the buzz of the engine, the ground below, all becoming part of a distant background. The shadow in the clouds played in his mind again and again. The first time he'd seen it he could dismiss it as perhaps a piece of debris or a wisp of smoke. Not this time. He had observed it for several seconds, even fired at it.

Did I shoot at a product of my imagination?

Manfred rested his free hand on his leather helmet, thinking of the scar beneath it. Could the wound be affecting his judgment? His perception of the world around him?

Maybe it will go away.

What if it did not? Would it get so bad he'd end up going insane like Förster and Private Schaffner? Would he also babble about red eyes?

But how could two people who do not know each other be terrified by the same thing? Manfred had asked Kretschmer that last night after he sedated Schaffner. The staff surgeon had two explanations. Coincidence or the onset of mass hysteria.

And what is the explanation for me seeing this . . . shadow thing? Twice?

He had no intention of asking Kretschmer. He feared the answer the man might give.

The base suddenly appeared in the distance. One by one the planes landed. Manfred rolled his Albatros D.V. up to its tent. He grabbed the side of his cockpit to pull himself out, but stayed put. Pain crushed his skull. A sagging feeling spread through his body. He worried he wouldn't be able to stand if he got out. And the damned nausea boiled in his stomach. He leaned forward, breathing out loudly. *Go away. Go away.*

"*Herr Rittmeister.* Are you all right?"

Manfred turned to his mechanic, Ewers. The young corporal stood beside the cockpit, biting his lip, eyes blazing with concern.

"*Ja. Ja,*" muttered Manfred. He cleared his throat, forcing strength back in his voice. "*Ja.* Just thinking, *Unteroffizier.*"

Much as he wanted to, he could no longer stay in the cockpit. Ewers had become a mother hen these past few days, asking if he was all right after every mission. He couldn't afford to worry the young man anymore, perhaps to the point he shared those worries with the other mechanics.

With a groan he could not suppress, Manfred pulled himself out of the cockpit and slid to the ground. His legs wobbled when he planted his feet. He braced a hand against the Albatros's body.

Ewers stared at him, fidgeting.

"I have several holes in my upper left wing," Manfred blurted, trying to head off any more worry from the mechanic. "Courtesy of a British gunner."

"I will get them patched up immediately. Hopefully you made the Englishman pay."

"Sadly, it was not a good day of hunting. For myself, anyway."

Manfred walked away, concentrating on each step, fighting to maintain his posture. He could not risk collapsing in front of his men.

Or throwing up. He swallowed to push down a stream of bile slithering up his throat.

He eyed his tent, Moritz sitting by the entrance wagging his tail. He had to lie down. Just for a few minutes. Then he'd be fine. Then he could tend to –

"*Herr Freiherr.*"

Manfred groaned and turned to find Kretschmer approaching.

"What is it, Doctor?"

"While you were away, Sergeant Holm led a patrol to try and find the rest of Private Schaffner's unit. They did. They recovered five bodies."

"What happened to them?"

Kretschmer let out a short sigh. "You should see for yourself."

Manfred frowned. Much as he wanted to rest, a commander could not shirk his responsibilities, no matter how unwell he felt.

He followed Kretschmer on rubbery legs, praying they did not give out.

They passed the hospital tent when a scream erupted from within. Förester.

"When are they going to send an ambulance to collect him?" snapped Manfred.

"I wish I could tell you." Kretschmer shrugged. "I imagine since his injuries are not life-threatening, higher command does not think him a priority."

He is my pilot, so that makes him a priority to me. He would submit another request for an ambulance personally. Perhaps a dispatch with the name Manfred von Richthofen on it would speed up the process.

He followed Kretschmer to a tent on the far side of the base. As they neared, his face wrinkled at the stench drifting from it. The sour odor of decomposing flesh and human waste mixed with the stink of copper. It was a smell Manfred knew well from his time as a cavalry officer in Russia and France.

The smell of death.

"You may want this." Kretschmer handed him a white surgical mask.

Manfred put it on and the two men entered the tent.

Five bodies lay on wooden tables covered in white sheets. The dead soldiers retrieved by Sergeant Holm and his men.

Manfred took a couple of steps forward, then stopped. The foul odor clung to the air. Even with his mask he could still smell it. Not just smell it, feel it burn through his nostrils and slide down his throat into his stomach. A stomach already swirling with post-flight nausea.

No. He gritted his teeth. *No, I cannot.*

His gut roiled.

I must . . .

Bile raced up his throat.

Not . . .

Manfred ripped off the mask and doubled over. Vomit exploded from his mouth. He drew a ragged breath, choked on the vile smell that surrounded the tent, and heaved again.

Coughing, he wiped his mouth and straightened up. Kretschmer stared at him, his eyes alight with concern.

Manfred scowled, fighting to not look away from the staff surgeon. The last thing he wanted was to throw up in front of anyone. What would that do to the morale, the confidence of those under him if their commander got sick in front of some dead bodies?

"I will not tell anyone of this," said Kretschmer.

Manfred simply nodded. "So what do you wish to show me?"

"It's their wounds. They are . . . well, not something I've come across before."

Brow furrowed, Manfred stepped closer to Kretschmer as the doctor pulled back the sheet of one of the soldiers. He was lean with a narrow, stubbly face. What should have been a gray tunic had turned blackish-red from dried blood, which had come from four gashes just below his chest.

Manfred cocked his head and stared at the wounds. His first thought was a knife. But the cuts looked too parallel, like they'd been done at the same time. What man could wield four knives at once?

"The others have similar wounds," said Kretschmer.

One by one he unveiled the other dead soldiers. Two had been slashed across their chest. Another across the face and throat, and the final one – which made Manfred grimace and set his stomach aflame with nausea – had been nearly disemboweled.

"This cannot be a knife." Manfred shook his head. "Or even a sword."

"I agree. Another thing I found in my examinations. None of these men had any bullet wounds."

Face scrunched, Manfred turned to the doctor. "How is that possible with all the shooting we heard?"

Kretschmer shrugged. "It would suggest these men were not fighting a British raiding party as we first suspected."

"Then who were they fighting?"

"I had considered wolves."

"Wolves?" Manfred drew his head back in disbelief. "I did not think there were any left in this region."

"True, they have been hunted to near extinction throughout most of Europe," said Kretschmer. "But there could be a few packs hidden away. Still, I do not think this is the work of wolves."

"Why not?"

"No one heard any howling last night, correct?"

"No," replied Manfred. "Just rifle fire."

"Mm-hmm." Kretschmer gave a brief nod. "And with all the shooting these men did, they would have hit some wolves. Yet Sergeant Holm and his men found no trace of dead wolves anywhere."

Manfred let out a long sigh. "So more mysteries for us to unravel."

"It appears so."

"Including how what happened to these men could be connected to what happened to *Leutnant* Förster."

"Who says it is?"

Manfred snorted. "Are you still sticking with your story that it is a coincidence?"

Kretschmer glanced to the side. "Maybe imagining red eyes is a symptom of shellshock. Something to do with the color of blood, perhaps?"

Manfred cranked an eyebrow. "Do you truly believe that, Doctor?"

The other man's shoulders sagged. "I don't know. All I have are theories on a malady most in my profession have not even researched. But if it is not a product of fear-induced imagination, then what? There really is some . . . thing out there with red eyes that can attack our soldiers both on the ground and in the air?"

A chill went up Manfred's spine. All his muscles coiled, his mind recalling the moth-winged form he chased and shot at. Had it been real? And if so, what the hell was it?

He stared at Kretschmer. Maybe he should tell him. Surely an illusion could not last as long as that thing he encountered in the clouds. Maybe his story could help the doctor figure out what happened to Förster and Private Schaffner.

Or what if I did imagine the entire thing? The doctors who treated him during his convalescence thought he had returned to combat too soon. If Kretschmer thought he might be hallucinating, he could have him grounded.

"Herr Freiherr?" Kretschmer tilted his head. "Are you all right?"

Manfred sneered. He was starting to hate that question. "I am just thinking."

Kretschmer nodded slightly, not looking entirely satisfied with the response.

"So we have no idea who or what killed these men and sent *Leutnant* Förster and Private Schaffner into hysterics?" Manfred spoke before the doctor became too suspicious.

"Not as of yet." Kretschmer shook his head. "Right now, there is only one thing I know for sure."

"What is that, Doctor?"

"Something very out of the ordinary is happening here."

SIX

Anger surged through Manfred, begging for release. He clamped his jaw tight as he stared across the desk at the target of his fury, *Leutnant* Erhardt.

He took a deep breath, hoping it would steady him. It did, only just. Why this urge to scream at the pilot? Yes, he had violated one of the main tenets of *Dicta Boelcke* by abandoning his wingman. Such acts frustrated him to no end. But he pointed out those mistakes in a calm yet stern tone. He had never been one to yell, considered it counter-productive to teaching and morale.

So why did he want to do it now? Bad enough he might be having hallucinations of moth-winged shadows in the sky, now he had to fight to bring his emotions under control.

He rubbed his bandaged scalp. Ever since the head wound, things seemed . . . different.

It will pass. It just needs time to heal.

He swallowed back his anger and refocused on Erhardt. "While you did shoot down your prey, you were almost shot down yourself." Manfred did not speak too loudly, but his voice was similar to that of a father lecturing an errant son.

"You never saw that Bristol behind you, did you?" he continued.

"No, sir." Erhardt frowned.

"And you would not have seen him until the moment he riddled your plane with bullets. That is what happens when you leave your wingman. You were extremely lucky I was able to drive off that British fighter. The next time you may not be so lucky, or perhaps *I* may not be so lucky if you are not there to support me."

Erhardt grimaced and shifted in his seat. "But *Herr* . . . *Herr Rittmeister.* That Bristol was going for you. I did chase it off."

"But you chased it too long and too far away from me," countered Manfred. "This unit is most successful when we fight as a team, not as individuals. If you wish to remain a part of this wing, you will do well to remember that."

"Er, *ja, Herr Rittmeister.*" Erhardt dipped his chin, resembling a scolded child.

"In the future, do not stray far from your wingman, and do not become obsessed with your target. Mistakes like that will eventually catch up to you and get you, and perhaps one of your squadronmates, killed. Is that understood?"

"Ja, Herr Rittmeister."

"Good. Return to your tent, go over *Dicta Boelcke* again, and apply it to your next mission."

"Ja, Herr Rittmeister." Erhardt got up, saluted, and left the tent.

Manfred watched him go. It seemed the pilot had taken his words to heart. *Which he probably would not if I had yelled at him.*

He leaned back in his chair and sighed. If only Erhardt had followed the *Dicta* during the fight with the Bristols. Maybe he would have seen the flying shadow, too, confirmed that it was real.

Or maybe Erhardt would not have seen it. That would just make Manfred more concerned for his sanity than he already was.

***

At first light the next morning, Manfred dispatched a messenger to 4th Army Headquarters requesting an ambulance for Förster and Schaffner. Hopefully one would be here soon. Along with giving whatever aid they could to the two men, Förster's screaming bouts could not be helpful to the confidence of his pups. How many wondered if they might crack like the *Leutnant?*

He should also have Dr. Kretschmer organize a burial detail for the soldiers killed by . . . whatever killed them. It wouldn't be long before the stench of decomposing flesh permeated the base. An honor guard would also be needed. He'd leave that to Sergeant Holm.

I will see to it after breakfast.

The morning meal had not been too bad. Manfred and the others had a decent serving of eggs and bread, but only one slice of bacon. As usual, several men complained about the portions and the blandness of the food. Manfred did not even try to silence them. Why bother? Soldiers had complained about the quality of their food since the Greek city-states warred with one another.

Once finished, Manfred returned his empty plate to the serving line. He'd just set it down when a short soldier with goggles pushed atop his leather helmet strode into the tent.

"Excuse me. I've been sent by Fourth Army Headquarters with a message for *Rittmeister* von Richthofen. Where might I find him?"

"I am here." He turned to the messenger.

The young soldier's eyes bulged in awe, as though unable to believe he was in the presence of the famed Red Fighter Pilot.

"*Herr . . . Herr Rittmeister.* It is an honor."

"*Danke.*" He nodded. "Now, the message?"

"Oh, yes sir." He reached into his leather pouch and pulled out several rolled up pieces of paper. "From General von Armin."

Several of the pilots in the mess turned to one another. Stunned whispers were exchanged. No surprise when their commanding officer had just received a message from the man in charge of the entire 4th Army.

"What do you think the general wants?" asked Voss.

"Obviously it's something important," said von Döring. "Generals are not in the habit of sending a message to simply say hello."

Manfred unfolded the first paper and read it over. He let out a quick breath and said, "You are right, von Döring."

"Sir?" the pilot's brow furrowed.

He held the paper out in front of him. "This is important."

SEVEN

The mess tent became a briefing hall. Within an hour, easels and chalkboards were brought in, some displaying maps provided by the messenger. One image showed a cigar-shaped biplane with a fin-like tail.

Manfred, pointer in hand, surveyed the assembled pilots. All looked to him, ready to begin.

"Per General von Armin, we are to conduct a search for an Albatros C-Three that failed to return to base yesterday."

Many of the pilots turned to one another with faces scrunched in either surprise or disbelief.

"Why does he want us to search for a reconnaissance plane?" Voss wondered aloud.

"We are a fighter squadron," von Döring jumped in. "I thought our priority was to establish air superiority over Flanders."

"It could have to do with our experience," Manfred answered. "Or we could be the closest aerial unit able to carry out this mission. There is also the fact that he is a general, which means he has the prerogative to assign any unit to any mission he wants."

Several pilots responded with resigned nods.

Manfred continued. "The C-Three was tasked with photographing the British lines just north of Hooge, four kilometers east of Ypres." He tapped the map with his pointer. "General von Armin fears that if the plane went down and was not completely destroyed, the British could recover the camera and any pictures with it. Even though he did not mention in the dispatch why he is so adamant for us to find this plane, I do have my theories. One, we are planning an offensive in that part of the Ypres Salient and those photographs are vital to its execution. Two, the British are planning their own offensive and those photographs are needed to aid with the preparation of our defenses. Whatever the case, should those photographs fall into enemy hands, it could jeopardize our battles plans, be they offensive or defensive."

Manfred moved his pointer to several red lines extending across West Flanders toward Ypres near the French border. "*Jastas* Ten and Eleven will conduct the sweep for the C-Three. We will fly two-plane formations at one-kilometer intervals from an altitude of eleven hundred meters. *Jasta* Four will trail us flying at three thousand meters to protect us from any enemy fighters."

He turned back to his men, his lips pressing together for a moment. "I am afraid enemy planes will not be our only concern. The meteorological service expects rain across Flanders today."

Several pilots groaned and shook their heads. Manfred didn't, but was just as worried. Flying through foul weather was not merely miserable, but could be as deadly as any machine gun.

"Does the general still want us to fly if it rains?" asked Erhardt, a slight treble of apprehension in his voice.

"His orders do not say otherwise, so it appears we must search even if the weather worsens."

More groans emitted from the pilots, along with frowns, scowls, and shifting in seats.

Gude raised his hand. "What do we do if we find the C-Three?"

"The general said if we are able to land nearby, we are to do so and make for the plane. We are then to recover the camera and burn the plane to prevent the enemy from examining it."

"And if we cannot land?" asked Heldmann.

"Then we are to strafe the plane and hopefully set it ablaze."

Adam let out a heavy sigh. "The one problem I see is if the plane went down among the British lines."

Manfred frowned slightly. "Our orders are still the same. If we come across the plane and it is relatively intact, we are to strafe it. General von Armin made it clear. The British cannot get hold of that camera."

"What if they already got hold of it?" Voss flicked his hand to the side for emphasis. "Then this whole mission is pointless."

Manfred cast his gaze down for a moment. Voss's concerns mirrored his. Also, he did not like flying at such a low altitude. It exposed them to ground fire.

I do not like this mission, period. He considered it a waste of resources. This was a job for other reconnaissance planes, not fighter squadrons. Von Döring was correct. All their focus should be on knocking the Entente out of the skies, especially when everything pointed to a major battle on the horizon.

In the end, his feelings did not matter. None of their feelings mattered to General von Armin.

Manfred straightened. "I know this is not the sort of assignment we would expect to be given. But these are our orders, from no less than the commanding general of the Fourth Army. We will carry them out, and we will be successful. Is that understood?"

"Ja, Herr Rittmeister," the pilots said in unison. Though in more than a few cases, their tone was not as enthused as Manfred would have liked.

Soldiers are not required to be happy about orders. They are required to obey them.

The men looked over their assignments and flight paths before Manfred ushered them out of the tent. He wanted to get in the air as soon as possible and finish the search before the storm came.

He studied the sky on the way to his plane. It was blue with several thick cottony clouds overhead. It seemed a nice day, at least so far. Might that change in an hour or two?

Manfred strode toward his red Albatros D.V. His pace slowed, ultimately stopping a few meters from the machine. His cheek twitched. He eyed the sky again, thinking of the moth-winged shadow. What if he saw it again? Would it prove he was losing his mind? What if it got so bad it affected his judgment in battle? To the point it cost the lives of his pups?

No. I will not let that happen. I can fight through this. I will get better.

What if he didn't? What if he could not hide this affliction from his men? Or worse, his superiors?

His jaw stiffened. He knew the answer to that. They would ground him. Not just ground him. His military career would be over. The family name of von Richthofen would be disgraced. Being wounded from a bullet or shrapnel was one thing. That was part of war. To be removed from duty because you lost your sanity . . . that would be a sign of weakness. Perhaps cowardice. The newspapers and films had built up a mythic image of him as Germany's greatest fighter pilot, its greatest hero. What would it do for the morale of both the Army and the civilian population if *Der Rote Kampfflieger* was declared a raving madman?

He caught sight of Corporal Ewers staring at him, head tilted, his concern evident.

Dammit. He pulled on his gloves, pretending to adjust them, then did the same with his helmet. Manfred hoped it fooled the mechanic into thinking he was just making sure his uniform fit properly. He did not want to hear the man ask if he was all right.

Manfred covered the remaining distance to his Albatros. *"Guten morgen, Unteroffizier."*

"Guten morgen, Herr Rittmeister. Are you --"

"Let's see how you did with those repairs." Manfred cut him off. He leaned under the left wing, then nodded. "Good as new. Excellent work as usual."

"Thank you, sir," replied Ewers. Only a trace of a smile formed on the young man's lips. Concern lingered in his eyes.

"Hopefully we will be back soon." Manfred climbed into the cockpit. "Rain is coming. I do not want to be caught in a storm and give you more damage to repair."

"I appreciate that, *Herr Rittmeister.* Good hunting."

Manfred nodded to the corporal. He then cranked the starting magneto on his left-hand side. The Mercedes D.III engine grumbled to life and the propellers spun. The RPMs steadily rose. The plane moved along the grassy field. Manfred tensed as he stared at the sky. Already he could see a smudge of gray in the distance. His worry spiked.

Not so much for the storm as it did the thought of encountering the flying shadow again.

EIGHT

Manfred's cheek twitched nervously as he eyed the horizon. The sky had grown darker since he and the rest of the wing took off. His mouth went dry, imagining wind and rain battering his plane and those of his pups.

And he had to worry about taking ground fire from the Entente, and whether or not he'd see that . . . thing again.

Worst of all, they might never find that damned reconnaissance plane, meaning some of his pilots could die for nothing.

The thought made Manfred glare behind his goggles as the wind buffeted him. He looked below. No sign of the Albatros C.III. He checked left and right. The other planes maintained their heading.

To his left, Manfred spotted a village. Rather, what remained of one. From his vantage point it looked like a big lump of black and gray, broken up by the remnants of stone buildings that somehow still stood. He then glanced around the sky, checking the storm clouds, his other planes, and watching out for any enemy aircraft. Then he went back to searching the ground, letting out a harsh grunt as he did. General von Armin should have assigned two-man observation planes for this mission, not single-seat fighters. The extra pair of eyes would have helped immensely.

The greens and browns of the landscape soon vanished. The ground became blackened and riddled with craters from past bombardments. He grimaced, remembering his first few months of the war. Down there, as a cavalry officer. The crash of artillery, seemingly never-ending. All he could do was huddle in a hole and pray one of those shells did not land on his head.

His jaw clenched. He was fortunate he no longer had to endure that hell. Not that it was any safer in the air. At least up here he could shoot back at his foe. He couldn't do that to an artilleryman blasting at him from several kilometers away. The battles up here also had a sense of honor to them. Dueling one-on-one with another flier.

He may have started out on horseback, but here in the sky was where he was meant to be. Not in supply, as the army wished to put him when cavalry proved useless on this new battlefield.

Manfred sneered at the thought. A Prussian was not meant to dole out blankets and cooking pots during a war. A Prussian was meant to fight.

But what if they will not let me? With his free hand he rubbed the top of his helmet, picturing the scar beneath it. What if he could not keep himself from collapsing or throwing up every time he landed? What if he continued to hallucinate?

What would become of him if he could not fly and fight?

Manfred shut his eyes. *Focus. You are on a mission.*

He gazed upon the torn landscape below. He found no trace of the C.III. Another check of the sky ahead of him showed the dark clouds getting closer.

He turned his head left to right. None of the other planes broke formation.

Another check below. More craters, more scorched earth. No C.III.

Manfred flew a few more kilometers. Another village appeared, this one not as ravaged as the other he'd seen. Green and brown reappeared, though marred by the occasional crater. This part of Flanders, it seemed, had avoided the worst of the war.

The village soon vanished behind him. The storm clouds, however, did not. They seemed ready to surround his planes. Manfred's shoulders tensed. It wouldn't be long before rain pelted them.

He looked down again . . . and sucked in a quick breath. He leaned forward in the cockpit.

The Albatros C.III lay below. It appeared to lean to one side, its left wing damaged. Otherwise, it remained remarkably intact.

Manfred nodded, silently praising the pilot for keeping his plane in such good condition. He wondered if the pilot was anywhere nearby. He would love to hear about this landing. More likely, both he and the observer had left the plane to avoid capture by the British.

But wouldn't they have destroyed their plane before leaving? His brow furrowed, concern slithering through him.

Manfred waggled the wings of his Albatros. He turned to Erhardt's plane, raised a hand, and stabbed a finger toward the ground. Erhardt rocked his fighter side to side, indicating he saw the crashed C.III.

Manfred gazed below. He spotted what seemed a relatively flat stretch of grass a little over a kilometer south of the downed reconnaissance plane. He signaled for Erhardt to follow him down.

The two planes swung toward the field, Manfred leading the way. He flew a couple of kilometers before turning and lining up his approach. The ground steadily rose. Manfred clenched his teeth, praying there were no ruts in the field that could damage his Albatros. He had no desire to hike through the frontlines to reach his base.

Or worse, get captured.

Manfred stiffened just before the wheels thumped against the earth. The plane bounced, jostling him in the cockpit. He slowed his speed and came to a stop.

Pain slammed into his skull. Manfred shut his eyes tight and leaned forward. He pressed a hand against the instrument panel, letting out a long groan.

Go away. Go away, he ordered the pain. It hung on, threatening to crush his brain.

Manfred bared his teeth. He couldn't afford this right now. Dammit, when was this agony going to end? When would he be able to fly without feeling like his head was about to explode?

Stop. Please stop.

The tight grip on his brain eased. Manfred managed to peel back his eyelids. He let out the breath he'd been holding for a long time. His muscles uncoiled as he leaned back in his cockpit. After sucking down a few more breaths, and with his headache lessening, he climbed out of the plane.

His boots just hit the ground when he noticed movement out the corner of his eye. Manfred whipped around. Dizziness rushed through his head. He pressed a hand against the side of his Albatros. His other moved toward his holster.

"Herr Rittmeister? Are you all right?"

Manfred blinked. His vision cleared. Erhardt approached. He hadn't even realized his wingman had already landed.

"Fine," he muttered before turning away from him.

Bending down, Manfred examined his plane's undercarriage. He found no damage to the wheels or struts.

"Did you check your plane?" he asked Erhardt without looking at him.

"Yes, sir. All in good condition."

"Very well." Manfred rose, slower than he would normally. He did not want to risk another dizzy spell. "Let's get moving."

The pair set off. Manfred gave one last look at his Albatros D.V and winced. He hated leaving his plane unguarded. That, however, couldn't be helped. At least no one besides them seemed to be around this section of the front.

That we know of. He wasn't even sure which side of the line he had landed on.

Manfred drew his Luger, even though all it did was give him a false sense of security. The pistol would be nearly useless against enemy infantrymen with rifles. There was really just one reason a pilot carried a Luger. To use on himself in case his plane caught fire in mid-air. Better to end things quickly with a bullet to the head than suffer the agony of burning to death.

He and Erhardt trekked across the field. A soft rumble echoed in the distance. Too sustained to be thunder. It had to be artillery. Thankfully, far enough away to not threaten them.

Manfred constantly checked his surroundings. No patrols, German or Entente, were in sight. The storm clouds, however, crept ever closer. The veins in his neck stuck out. There was no chance of getting back to the air base without getting soaked.

Manfred quickened his pace.

It didn't take long before he spotted the downed C.III in the distance. Manfred broke into a jog, Erhardt following his lead. As they got closer, Manfred saw a shape on the ground. A man. Likely the pilot or the observer. He lay face down, unmoving.

"Erhardt, check the plane," ordered Manfred as he approached the fallen airman. Flies buzzed around the body. He grimaced at the stench permeating the air. A sickly, sour smell of decomposing flesh and bodily waste.

Manfred stood over the dead flier, his eyes tracing the dark pools stretching from the corpse. Had he been shot in the air and managed to crash-land before dying?

That's when he caught sight of the Luger resting about a meter from the pilot's hand. Could he have been shot on the ground? Manfred snorted at the thought. If an enemy patrol had come across him, they must have searched the C.III and taken the camera.

I hope that is not the case. He bent down, grabbed the pilot's jacket, and rolled him over.

Manfred's breath caught in his throat. His eyes widened.

Four deep slashes cut across the man's chest.

NINE

Manfred could not take his eyes off the gashes that stretched across the pilot's chest. His throat tightened as his mind recalled similar wounds on the men of Private Schaffner's patrol. It appeared whatever killed them also killed the C.III pilot.

Except Schaffner's patrol was dozens of kilometers from here. Just what the hell was –

"Sir," Erhardt called out from near the tail of the plane.

"What is it?"

Erhardt paused. "I think you should come see for yourself."

Manfred made his way over to him, glimpsing a hump sticking out from the rear seat. The observer, likely dead as well.

He halted beside the other pilot, eyeing the plane's shredded tail section.

"Look." Erhardt pointed. "These are not bullet holes."

Manfred stiffened as he gazed at the four jagged tears along the rear half of the fuselage. Erhardt was right. No bullets had done this.

Slowly, Manfred looked back at the dead pilot. "It looks similar to the injuries he has." Grimacing, he turned to Erhardt. "Him and the soldiers from that patrol near our base."

Erhardt scrunched his face, eyeing the slashes on the plane. "What could be doing this?"

"I don't know." Manfred inhaled deeply, taking in the odor of spilt fuel.

A rumble shook the darkening sky above. He looked up, jaw tight. That hadn't been distant artillery, but thunder. Almost on top of them.

"We do not have time to figure it out. Collect the pilot's tag and any documents he may have on him." Manfred nodded to the corpse. "I will see if the camera is still on board."

"*Ja, Herr Rittmeister.*"

Manfred strode up to the observer's seat while Erhardt headed over to the pilot. He placed his hands under the man's shoulders and pushed him back against the side of his seat. He held his breath when he saw the bloody slashes across the observer's neck and chest.

His shoulders slumped for a second. That was all the time he had to mourn his fellow flier. Manfred yanked the ID tag from around the man's neck and checked around the seat.

He breathed a sigh of relief when he saw the boxy camera laying near the observer's feet. It must have tumbled out of his hands when he died.

Manfred reached in, a raindrop pricking the back of his neck, and pulled out the camera and examined it. He let out a second sigh of relief when he determined it was intact.

"We actually found it." Erhardt beamed as he walked up to him. "I prayed we would." He tapped the Bible underneath his flight jacket.

"And the Lord answered your prayers, *Leutnant.*" He gave Erhardt a brief grin. "All of our prayers. I feared this mission would prove fruitless."

He set down the camera, and with Erhardt's help, dragged the observer out of the plane. They laid him beside the pilot as another raindrop plopped down on Manfred's shoulder. He gazed at the dark gray clouds hanging overhead. Another drop hit his face.

Manfred stared at the camera in his hands, his lips compressed in worry. Would rain seep into the camera, ruining the film inside? They had been beyond fortunate to find it undamaged. He could not allow this storm to render their mission pointless.

He handed the camera to Erhardt and bent over the body of the observer. Manfred grasped the man's jacket, then hesitated. It felt disrespectful to steal clothes from a dead man. But he had no choice.

This is in service to Germany. He would understand.

Manfred removed the flight jacket from the deceased observer and wrapped it around the camera, tying the sleeves together to ensure it stayed covered. That should protect it from the rain.

"Should we bury them?" Erhardt looked at the C.III crew.

Manfred frowned and shook his head. "No. We have no shovels, and even if we did, there is no time. All we can do is hope a patrol comes by and gives them a proper burial."

"What if it's a British patrol that finds them?" A slight edge crept into Erhardt's tone.

"Then we hope they show some honor and bury them. Enemy or not, no soldier deserves to be left to rot on a battlefield."

Erhardt's mouth formed a tight line. His head bobbed slightly from side-to-side, as if trying to figure out whether or not to agree with his commander.

Manfred did not have time for Erhardt to make up his mind. "Search the observer for any documents. I'll set the plane on fire, then we need to go."

"Yes, sir."

Erhardt rummaged through the observer's flightsuit while Manfred set down the jacket covering the camera and took out his flare gun. He aimed for the rear of the C.III and fired. The spilled fuel ignited. Flames licked the plane's ruined tail section.

A few more raindrops splashed on him. Manfred scowled. He hoped the flames consumed the aircraft before it stormed in earnest.

Picking up the camera, he walked over to Erhardt, who stood clutching a sheaf of papers. "What did you find?"

"Maps. Similar to the ones I found on the pilot. The observer also had a notebook." Erhardt grinned. "Now the British won't get these or the camera. God was definitely with us today."

"Make sure to offer him another prayer to get us back to base sa --"

A crack echoed through the air. Something snapped past them.

Manfred threw himself onto his stomach. So did Erhardt. Though it had been a couple of years since either of them fought on the ground, some of their soldierly instincts remained. Like ducking when bullets began flying.

He lifted his head as much as he dared, staring past the burning C.III. Manfred saw nothing, though the smoke did obscure his vision. He checked to the right.

There! About a hundred meters distant. Six figures jogged toward them. The soup bowl-shaped helmets left no question who they were.

The British.

Dammit. Manfred clenched his teeth, fingers pressing against the jacket containing the camera. There was no way he could let it fall into their hands.

Or him. The newspapers and films back home had built him into a national hero. He couldn't imagine the propaganda coup for the British if they captured *Der Rote Kampfflieger.*

He looked over his shoulder at Erhardt. "We have to move."

The pilot's face stiffened as he nodded.

"Go!"

Manfred scrambled to his feet, camera in hand. He bent at the waist and ran in a zigzag pattern. Erhardt imitated him.

More shots erupted behind them. Two bullets cracked past. Two more sent up small spouts of dirt and grass nearby.

He and Erhardt kept running. Left and right, left and right, always changing the number of steps before their next turn. Anything to try and throw off the aim of the British.

The rain fell at a steady pace. Manfred's lungs burned. He took one deep breath after another and kept pumping his legs. He couldn't afford to stop until he reached his plane.

Manfred checked behind him. Erhardt kept pace. So did the British. They sprinted past the C.III, flames now whipping from one end of the aircraft to the other. He spared a moment for a quick grin. At least the British would not get their hands on it to study.

Another shot made him run faster.

"Erhardt. Send up a flare."

"Ja," he replied breathlessly. Still running, he grabbed the flare gun hanging from his belt. He dropped to one knee, raised the gun, and fired. The glowing red projectile had barely left the barrel when he started running again. A bullet kicked up dirt a couple of meters from him.

Manfred glanced at the flare as it ascended to the sky to alert his *jasta.* Hopefully they could reach them in time to deal with the British.

He pounded up a small rise. In the distance he could make out his red Albatros D.V and Erhardt's black and orange Albatros DIII. He ignored the burning of his lungs and legs and ran faster. Left, then right, then left again. Almost to his plane. Almost to safe—

Manfred slipped on the wet grass. He toppled onto his side. The camera tumbled across the ground.

"Dammit!" he snapped.

"Sir!" Erhardt dashed toward him.

Manfred began to push himself to his feet when Erhardt grabbed his arm and yanked him up the rest of the way.

A spout of dirt shot up near the camera. Manfred's chest tightened. That was too close.

He snatched up the camera and checked behind him. The British were less than fifty meters away.

He took off running, Erhardt on his heels. Both continued their zigzag pattern. Another rifle cracked. A small breeze whipped past Manfred's right arm. Cold fear pierced his chest. A couple of inches to the left and it would have been the end of him.

But he couldn't dwell on it. Their planes were close. All that mattered was getting to them and taking off with the camera intact.

Another round snapped past. Another. The rain came down harder, slapping his face. He blinked the excess water from his eyes, still running. To the left. To the right. Another bullet zipped over his head.

He reached his Albatros and chucked the camera into the cockpit before climbing in. Manfred ignored the damp seat under him and cranked the magneto. He used his feet to shove the jacket-covered camera beneath the seat as the engine grumbled to life and the propellers kicked in.

Past the spinning blades, he saw the British patrol charging for him. Less than thirty meters away. One raised his rifle and fired. He felt a thud ripple through his plane.

Manfred's heart hammered. He checked the RPMs. Still not enough for him to take off.

Muscles tensed, he looked at the approaching British. *I'm not going to –*

Something flashed to his right. Manfred swung his head and looked up. A fighter barreled through the rain. A Pfalz D.III.

He smiled. It was Heldmann.

A line of bullets ripped up the ground between him and the British. The enemy soldiers dove on their stomachs as Heldmann's plane buzzed over them. It did not appear he had hit any of the Brits. But it would keep their heads down. That was all Manfred needed.

He swung the Albatros D.V. around. The engine roared as he raced along the field. The plane, and him, bounced with every rut it hit, and there were quite a few.

Hold together. Hold together. He clenched his teeth as he pulled back on the stick.

The Albatros rose. Manfred gritted his teeth as the rain battered him, rivulets dribbling down his goggles. He wiped them away with a gloved hand. Just ahead of him was Erhardt. He came up alongside the other pilot, who gave him a short wave. Manfred returned the gesture, then gazed around at the sky. Gray clouds and a curtain of rain reduced their visibility. Instead of seeing for dozens of kilometers like on a sunny day, he could barely see a kilometer in front of him.

He signaled Erhardt that he would take the lead to guide them back to base. Manfred pulled his Albatros ahead. Not too far. In these conditions it would be too easy to lose sight of each other.

Manfred checked his compass. Northeast was where they needed to go, and that's where the needle pointed.

A village passed below. Which one he wasn't sure. He just kept the nose pointed northeast, giving quick glances at his watch, trying to gauge how far he had to go to return to base.

A gust of wind knocked his plane left. Panic washed through him. He subdued it and brought the fighter under control. A check over his shoulder revealed Erhardt remained on his tail.

Tremors gripped him. Soaked to the bone, the wind and rain sent an icy cold burrowing into his bones. His teeth chattered.

Concentrate on flying.

His shivering made that nearly impossible. His thoughts turned to standing in front of a stove, draped in a blanket, holding a mug of hot coffee. To be warm again.

You'll be warm if you crash this plane and it explodes into flames. Now focus!

Manfred stared straight ahead, clenching his teeth, barely able to keep them from chattering. He checked his compass. He was still on the right heading. Next, he looked at the fuel gauge. More than enough to get back to base.

So long as I don't get blown off course.

He glanced around the stormy sky, his stomach sinking. What were his pups going through? Would they all return to –

The Albatros dropped as if pushed down by a giant invisible hand. Manfred gasped and fought with the stick. The wind kept pushing him down.

Then suddenly he rose. Through rain-splattered goggles he saw something above him. Black and orange, with wings.

Erhardt's plane!

Manfred slammed the stick left. He glimpsed the D.III's wing just meters off his right side. Erhardt snapped his fighter right.

Manfred straightened out his Albatros. Erhardt also leveled out. Manfred took up position in front of the other pilot, sucking down quick, shaky breaths. He continued to shiver, this time because of more than the cold.

Just a few meters and I'd be dead.

He grimaced. *So would Erhardt.*

Manfred scowled at the rain around him. At least with another plane he could shoot at it. Thunderstorms cared nothing about bullets. All he could do was keep flying.

And keep praying.

The rain kept coming down. The gray clouds blotted out the horizon. Manfred searched for any landmarks. The rain made it difficult to pick out familiar sections of earth.

His brow furrowed. The patter of the rain against him and his plane seemed to soften. Manfred looked up. Could it be . . .

The rain started to lighten. Breaks formed in the clouds. Slowly his muscles uncoiled. Had they made it through the worst of the storm?

The sky dueled between blue and gray. The shower became a drizzle. Visibility expanded to several kilometers. Manfred could finally pick out distinct features below. The river to the north. The village off to his left with its prominent church steeple.

He held his breath when he realized it. They were off course by at least forty kilometers!

Manfred looked at his fuel gauge and swallowed. He might have enough to get back to base.

He *might.*

Manfred banked the Albatros to the north, staring back at Erhardt, who followed. He grimaced. If he was low on fuel, the same had to be true for his wingman.

You can make it. Please make it.

The drizzle continued, but Manfred ignored it. His eyes flickered between the sky and the fuel gauge, the needle sinking ever lower. He examined the green and brown patchwork of fields below him, just in case he needed to make a sudden landing. He hoped it wouldn't come to that.

The airfield appeared in the distance. A long sigh flowed from Manfred's mouth. The needle on his fuel gauge kissed the empty mark. He should have just enough to make it.

Half-a-kilometer from the field, the engine coughed, then died. His fuel had run out.

Manfred's chest tightened. The Albatros dropped toward the ground.

I should still have enough momentum. He pulled back on the stick, trying to keep the nose up and glide in. The sandbag wall grew before him. Manfred clenched his legs together. Would he hit –

The wheels just missed the sandbags. A quake rocked him as the Albatros's wheels thumped against the ground. With the engine dead, all Manfred could do was wait for the plane to roll to a stop. When it did, he climbed out of the drenched cockpit.

Dizziness swept over him. He pressed both hands against the fuselage to steady himself.

"Herr Rittmeister." He heard the concerned call of *Unteroffizier* Ewers.

Manfred pushed himself away from his plane, concentrating on his shaky legs. Only sheer willpower kept him from collapsing.

Ewers rounded the plane, along with a few other mechanics. "We were worried when we saw the propellers stop spinning. I should have known you'd find a way to . . ."

His voice trailed off, his eyes bulging. *"Mein Gott.* You're soaked, *Herr Rittmeister."*

"I know. Just move my plane to the side. We need to make room for the other planes coming in."

"Yes, sir." Ewers called over more mechanics to help. They just started pushing the Albatros when Erhardt's fighter touched down, his propeller still going. It seemed his wingman had a bit more fuel than he did.

"You should go to your tent and get warm, *Herr Rittmeister*," Ewers said with pleading eyes. "I will bring you some hot coffee."

Manfred couldn't help but grin. Ewers, always acting more like a doting mother than a father. "I will do as you order, *Herr Unteroffizier.*"

He took a couple of steps before back turning to Ewers. "There is a camera in my cockpit. It is wrapped in a jacket under the seat." He hoped that kept it, at the very least, relatively dry. "Make sure it is put in a safe place."

Ewers acknowledged the order as Manfred made for his tent. Once inside, he stripped off his damp clothes, put on fresh ones, then lit a fire in his small stove. Next, he lay on his cot and wrapped the blankets around him. He hated to be in bed while there were still things to do, but he had to warm himself up. He'd just gotten out of the hospital after recovering from his head wound. He did not want to go back with pneumonia.

Manfred found himself drifting off, but fought to stay awake. He thought about the C.III, the slash marks on the crewmen and the fuselage. Manfred tried to imagine how the fliers and aircraft met their fate.

They must have been hit by gunfire, whether from the air or the ground he could not say. Somehow the pilot brought the C.III down relatively intact. A remarkable feat of flying. A shame he could not celebrate it as whatever attacked Private Schaffner's patrol also set upon them. But who or what could be responsible?

He forced his mind to work through the possible events again. Shot down, survived the crash, killed by whatever had slaughtered the patrol. A group of soldiers and a reconnaissance crew with similar injuries, but so far apart from one another.

Manfred visualized the reconnaissance plane in his head, focusing on the damage, trying to imagine what could have befallen the crew after it was shot down.

His eyes snapped open. He stared at the top of the tent, recalling what Erhardt had told him.

The C.III did not have any bullet holes in it. It had not been shot down.

Then that would mean . . . No. Impossible.

But Manfred could not think of any other explanation. Whatever had slashed the plane must have done it while it was in the air!

TEN

Manfred lined up on the SPAD when the biplane banked left.

"Dammit," he muttered, giving chase. With the skies overcast, he was unable to dive out of the sun and surprise his prey. That gave the British pilot the chance to spot him.

He glanced around to make sure no enemy was approaching. He caught Erhardt's D.III pursuing the SPAD's wingman.

Good job. Manfred looked back at the SPAD in front of him. It twisted right. He turned with it, putting his sights on the left wing. The distance was a bit farther than he would have liked, but he still thumbed the fire buttons. Pieces of wood and canvass tumbled away from the aircraft. It banked left.

Manfred stayed with him. Another burst should rip apart the wing enough to bring it down. Kill fifty-eight. His first since his injury.

He grimaced, thinking of the gash in his head. Manfred grunted, shrugging off the errant thought. He glanced to the left. Clear. A glance to the right.

Another SPAD bore down on him.

He pulled the stick back. The Albatros D.V shot up into the gray sky. A heaviness pressed down on his body. Normal for whenever he made tight turns or quick climbs. He bared his teeth, groaning, then nosed over. He sighted the SPAD that had come after him, the imbecile flying straight. Had to be a novice pilot, thought Manfred. Probably shocked over what looked to be an easy target avoiding his fire, now unsure what to do.

A fatal mistake.

Manfred angled his Albatros into a dive, bearing down on the enemy plane.

Another fighter beat him to the prize. From the colors and markings, it belonged to Heldmann. Orange flickers spat from the twin LMGs on its cowling. Large splinters jumped off the SPAD's fuselage. Smoke belched from its side. The enemy fighter rolled over and fell toward the ground.

Manfred nodded. He may have not gotten the kill, but at least one of his pups had knocked down a British plane.

He searched the skies for another potential target. Two other fighters went down in flames. He had no idea if they were German or British. Two more planes, both SPADs, retreated south toward their own lines. Probably low on fuel. More SPADs soon followed. Some of the aircraft from *Jagdgeschwader 1* also headed back to base. Manfred checked his fuel gauge. He had enough to return home, but not enough to continue the fight.

He turned east, checking that no British fighters were trying to jump him. All clear. Manfred scowled as he stared straight ahead. Once again, no kills. Yes, he had damaged one of the SPADs. But it would be out of action for a day or two at most, then return to the sky. Maybe to shoot down a German plane and kill its pilot. A destroyed enemy aircraft and dead enemy pilot could not do that.

Erhardt joined up with him and the two continued east. Manfred studied the sky. Still cloudy, but thankfully no rain. The storm he'd flown through three days ago had been one of the most harrowing experiences in his flying career. But he had returned to base.

Some of his pups had not been as fortunate. He hung his head, thinking of the four pilots that had been lost in the storm. Three other pilots had been forced to crash land, but managed to make their way back to the airfield.

Bad enough when the British shoot us down. We do not need a storm to do their work for them.

His face tightened as his anger rose. Whatever was on that camera he recovered and delivered to 4th Army Headquarters had better have been worth the price his unit paid.

One by one, his planes landed at the airfield. Manfred rolled his D.V up to its tent. He shut his eyes as a headache tore through his skull. Nausea punched his stomach. He drew one deep breath after another, fighting off the pain.

Footsteps approached. Most likely Ewers. He forced himself out of the cockpit, swallowing hard to keep from throwing up.

He planted his boots on the grass just as the mechanic stepped up to him.

"Herr Rittmeister."

"Ja," he mumbled, waiting for the mechanic's inevitable, "Are you all right?"

But Ewers said something completely different. "General von Armin is here to see you."

Manfred's eyes widened. He ignored his headache and nausea. "What?"

"He arrived about an hour ago. He is waiting in your tent. He said he wanted to see you as soon as you landed."

Manfred nodded, still not completely over the shock. *"Danke, Unteroffizier."*

He started toward his tent when Ewers called out, "What do you think he wants, sir?"

"I have no idea." Manfred gave a slight shrug. "I suppose I will find out soon enough."

"It must be something important."

"Well, I do not think the commanding general of the Fourth Army would come here on a social call."

Ewers nodded as Manfred set off. He eyed two cars parked near his tent. Four soldiers stood by them, clutching rifles. Von Armin's bodyguards, no doubt. They stood at attention as Manfred approached. He saluted them and ducked into the tent.

Seated by his desk was a slender man in a gray uniform adorned with medals, including an Iron Cross hanging from his collar. Thinning gray hair sat atop his angular face, which sported a pointed beard and bushy mustache.

"Herr General." Manfred stood at attention. "Captain von Richthofen reporting as ordered."

General der Infanterie Friedrich Bertram Sixt von Armin rose. His mustache twitched in a smile. *"Rittmeister* von Richthofen. Good to see you."

The old general, whose career went back to the Franco-Prussian War forty-seven years prior, extended a hand, which Manfred shook.

"I was admiring your collection while you were off fighting." Von Armin turned to the table containing the silver cups marking each of Manfred's fifty-seven victories. "Do you have another to add to it?"

"Unfortunately no, sir. Though some of my other pilots scored victories today."

"Good. Hopefully they will have more in the coming days. The British and their allies have begun their push on the Ypres Salient. Establishing dominance of the air is critical to our operations."

"Yes, sir. We shall achieve that."

"I have faith that you will, *Rittmeister* von Richthofen," said von Armin. "Which is why I have chosen your unit for a special task."

Manfred cocked an eyebrow. "Sir?"

Von Armin stepped over to the desk and opened his satchel. "The pictures from the camera you recovered have been developed. Along with some excellent images of the British lines near Hooge, there was one photograph that our analysts found . . . intriguing."

"How so, sir?"

The general pulled out a photo, pausing to stare at it. "It seems before he died, the observer of the C-Three took a picture of the aircraft that brought down his plane. Our analysts have never seen anything like it before. They believe it could be some secret weapon developed by Britain or France. Whatever it is, I feel *Jagdgeschwader* One has the skill and experience to find and destroy it."

Manfred nodded, concern gripping his mind. Was the general talking about some new type of fighter? What if it outperformed his planes?

His shoulders tightened. What if this gave the Entente the edge it needed to control the skies and, Heaven forbid, win this war?

Von Armin handed him the photo. Manfred looked at it.

His eyes bulged. The rest of his body froze in shock. *It cannot be.*

Staring back at him with glowing eyes was the image of the moth creature.

ELEVEN

Could I be hallucinating this, too?

Manfred rubbed the edges of the photograph. It was definitely real. The same had to be true with the image before him. General von Armin had seen it, so . . .

It does exist. All his muscles loosened as relief flooded through him. For weeks he feared he had been losing his grip on sanity. Now he had confirmation this creature was not just in his head. It was as real as the general standing before him.

Manfred gaped at the picture. The observer had captured a clear image of the creature before he died. It had a round head with no visible ears and broad shoulders. Part of its moth-like left wing was visible. And the eyes. Just two orbs that appeared to be glowing. With a black and white photograph, he had no way of knowing their true color.

But he could guess.

"Rittmeister?"

The general's voice sounded distant. Manfred did not acknowledge it.

"Rittmeister. What is wrong?" Von Armin was louder, more demanding.

Slowly, Manfred lifted his head.

Von Armin's face wrinkled, appearing both concerned and confused. "You look as though you have seen a ghost. What is wrong with you?"

Manfred let out a breath and lowered the photograph. What would the general's reaction be when –

"Rittmeister!" von Armin barked. "I asked you a question. Why has that picture disturbed you so?"

He faced the general, straightening. *"Herr General.* I wish to report that I have seen this . . . thing."

Now von Armin's eyes widened. His mouth hung open wordlessly for a moment. "Wh-What? You've seen this? When?"

"Twice during the past two weeks. The first time I only caught a glimpse of it. The second time was a few days later, during another aerial battle. I pursued it for a while and even fired at it, but missed."

Von Armin did not speak, just blinked, as though absorbing what Manfred had told him. Several seconds went by before the older man's eyes narrowed. "You saw this aircraft," he jabbed a finger at the photograph, "and did not report it to anyone?"

"It is not an aircraft, sir. That I am sure of. It is . . . a living creature."

Again, von Armin held him with a silent, appraising stare. Manfred studied the general's expression, looking for any sign the man thought him crazed.

Surprisingly, he found none.

"If I may explain, sir . . ."

The general nodded, and Manfred continued.

"As I said, I have seen this creature twice. But it was so . . . unbelievable, and no one else saw it. I . . ."

He bit his lip. The last thing he wanted was to admit any sort of weakness, especially to a general. But from the rigid look on von Armin's face, it seemed unwise to hide anything from the man.

"I feared that I might be imagining it." Manfred left out that he worried any possible hallucinations might have been due to his head wound. His doctors had tried to dissuade him from returning to combat. The less attention he brought to the injury, the better.

"If I had reported this creature, with no proof, would anyone have believed me?"

"You think no one would accept the word of *Der Rote Kampfflieger* without question?" asked von Armin.

"Truthfully, *Herr General*, would you have believed a story about a . . ." Manfred glanced down at the photograph, "flying . . . mothman, even from me, or would you have thought that person mad?"

Von Armin stared at the ground in apparent thought for a moment. "I will admit, *Rittmeister,* I would have likely ordered such a man thrown into an asylum. Even you."

He held out his hand toward the picture, which Manfred gave to him. Von Armin gazed at it and shook his head. "Some of our analysts have said this looks more like some monster than an airplane. Though they would not officially declare it to be something alive. Probably for the same reasons you gave. But with this photograph and your testimony, it seems we have discovered something . . . extraordinary."

"I do not believe I am the only one who has seen this creature."

Von Armin's eyebrows scrunched together. "Who else has seen it?"

"One of my pilots, *Leutnant* Förster. When we returned from a battle nearly two weeks ago, he went into hysterics when he landed, screaming about red eyes. There is also a private named Schaffner. His patrol was slaughtered not far from here. He went into a mad rant about red eyes."

"Where are they now?"

"An ambulance collected them yesterday and took them to an asylum back in Germany," Manfred told him. "But that is not all."

"What else is there?" asked von Armin.

"The soldiers from Private Schaffner's patrol who were killed had wounds similar to the damage myself and *Leutnant* Erhardt found on the tail section of the C-Three. Four horizontal lines. More like slash marks than bullet holes."

Von Armin pinched the bottom of his gray beard between his thumb and index finger and stared thoughtfully at the photograph. "So there is some sort of monster in this part of Flanders that can not only kill our soldiers on the ground, but bring down our airplanes."

"It would seem so, sir."

"Where could such a beast come from?"

Manfred shook his head. "I do not know, sir."

Von Armin snorted. "Not just where could such a thing come from, but how did the Entente find it and train it to attack us?"

"You think the British or the French are using it?" Manfred drew his head back a bit, unsure about the general's statement.

"Without a doubt. Has it not been the way of this war? Each side fielding some new weapon to gain an advantage. Gas, fighter planes, tanks. Now this, a living weapon."

The general's bushy mustache twitched. "Maybe the British or the French came across this monster in one of their African colonies. It is called the Dark Continent. Who knows what mysteries could be hidden there?"

Manfred stood with his hands behind his back, saying nothing. Nothing about this creature made him think of Africa or any of the other continents. Honestly, it looked like it did not come from anywhere on this world.

Not that he was about to contradict the general, especially when he had nothing to base it on except a feeling.

Von Armin continued, "Whatever it is, and wherever it comes from, it is a threat. Your unit has the best pilots in the Imperial Army. I can think of no other group that I would entrust with such a vital task."

"And what task is that, sir?" Manfred asked, though he could already guess the answer.

"You and your men are to find this – what did you call it – Mothman, and rid it from the skies."

TWELVE

Manfred looked out at the pilots gathered in the mess tent, then to the easel to his right. A cloth hung over it, covering the picture of the Mothman. His shoulders tensed as he gazed at the men of *Jagdgeschwader 1* again. How would they react when he showed it to them?

Once all the pilots were present, he began. "Men, we have been given a new mission. This comes directly from General von Armin."

Several pilots turned to one another. Eyebrows went up. A few smiled. The expressions held surprise and eagerness. Manfred figured they thought if the commanding general of the 4th Army had ordered this, it must be important. So important only this unit could carry it out. They must have felt honored.

Will they still feel that way when I show them our target?

"I will tell you," Manfred continued. "This mission is unlike any we have ever undertaken. In fact, I doubt any squadron, on either side of this war, has been tasked with such a mission."

There were more turns of the heads by his pups. More raised eyebrows, even some grins.

Manfred clutched the top of the cloth. He hesitated for a moment, then said, "This is our target."

He lifted the cloth. Everyone now had a clear view of the Mothman photograph.

Manfred studied the faces of his pilots. Many stared at the picture with wide eyes or mouths agape. A few reactions stood out to him. Erhardt appeared frozen in his seat. So did Voss of all people. Their gazes were fixated on the creature.

Another curious reaction came from Heldmann. The man seemed to shrink back, and shifted his eyes away from the photograph.

"This picture came from the camera *Leutnant* Erhardt and I recovered from the downed C-Three a few days ago. It is believed this . . . creature was responsible for bringing down that plane, as well as killing the patrol near our base."

Many of the pilots exchanged astounded looks. Erhardt still had his gaze fixed on the photograph. Voss sagged in his seat, looking down in . . . thought? Shock?

Von Döring raised a hand. "What . . . What exactly is it?"

"I wish I had an answer," replied Manfred. "The truth is, no one knows. The photographic interpreters first thought this was some sort of new airplane from the Entente. But it has been confirmed that this is some sort of . . . beast."

All eyes were now on him, even Erhardt's, though the young pilot looked . . . dazed?

"What we do know about this creature is that it can fly, apparently just as fast as our planes," said Manfred. "Its claws are sharp enough to slash through flesh and wood. *Leutnant* Erhardt and I found such slash marks on the tail section of the C-Three. Other than that, we know nothing else."

Manfred's gaze swept over the pilots. Erhardt's expression hadn't changed. Heldmann rested his chin on his hand, eyes down.

Then there was Voss. Manfred had expected some sort of quip from him, something to lighten the mood. But he stayed silent and . . . sullen?

He had expected astonishment from his pups. But with Erhardt, Voss, and Heldmann it was something else.

Wait. Manfred held his breath. *Could it be . . .?*

Manfred took two steps forward, tilting his chin up. "I have personally seen this creature twice."

All gazes swung his way. He caught soft gasps and others whispering, "What?"

"The first time was the day *Leutnant* Förster had his hysterical reaction. I only caught a glimpse of it. But the second time, just a few days after, I had a better view of it. I even chased it and shot at it before it vanished into a cloud bank."

Hands clasped behind his back, Manfred said, "Judging from some of your reactions, I suspect I am not the only one who has seen this Mothman."

His jaw clenched. He did not want to say what he was about to in front of these men who looked to him for leadership and inspiration. But given the threat this monster posed, he did not think it could be avoided.

"I did not speak of it to anyone . . . worried that everyone would think me mad."

Manfred glanced at the easel. "But this photograph is proof that this creature is real. It has killed our fellow soldiers and I believe it is the reason *Leutnant* Förster went out of his mind with fear. If we do not find this thing and kill it, more Germans will die. So I ask, anyone who has seen this monster, speak now. Whatever information you have could give us the advantage we need over it."

Voss' hand went up. So did Heldmann's. So did three more pilots.

To Manfred's surprise, Erhardt's hand stayed down.

"I saw it on my second flight with *Jasta* Ten," said Voss. "I was bearing down on a British fighter when it crossed in front of me. The damn thing threw off my aim. I told myself I imagined it, or that it was a shadow. It had to have been. A flying man with moth wings couldn't be real." He looked at the photograph. "I guess that picture proves me wrong."

Voss laughed. A brief, nervous one.

"I saw it two days ago," said Heldmann. "I was chasing a Canadian fighter and was checking behind me to make sure my tail was clear. That's when I saw it. It was only for a second or two, but it was that monster, I have no doubt."

He frowned. "I also kept what I saw to myself. How could anyone believe me? They'd think I was crazed or drunk. Or both. Either way, the Army would probably never let me fly again."

Manfred nodded. "You were not alone in those concerns."

Voss also nodded.

The other three pilots shared their stories. Like Voss and Heldmann, they had only seen the Mothman for a handful of seconds. Not long enough to learn anything more about its capabilities. Or more importantly, any weaknesses.

"So how are we supposed to find . . . *Herr* Mothman?" Voss jabbed a hand toward the photograph. "It could be anywhere in Belgium."

Manfred walked over to an easel on his left. This one displayed a map of the Flanders region, which encompassed the northern part of Belgium. He pointed to a circle that encompassed the Ypres Salient and extended to the city of Ghent.

"This is the area where all known Mothman attacks have occurred." He pointed to three red X markings. "*Leutnant* Förster's encounter, the Army patrol, and the C-Three. This is where we will concentrate our search."

"That is still a lot of territory to cover," noted von Döring. "And we have no guarantee this monster will be in the air the same time we are."

"Exactly." Manfred nodded. "Consider birds and insects. They fly, but not all the time. They have to nest at or near ground level. I believe the same is true for the Mothman. It must have a nest somewhere on the ground. I explained my theory to General von Armin, and he agreed with it. That is why he is assigning us two reconnaissance planes and a photographic interpretation unit. They should arrive tomorrow. I also hope to have replacement pilots here soon to make up for our recent losses."

His voice trailed off with the last few words. Several of the men lowered their heads.

"Also . . ." Manfred squared his shoulders as he spoke. "And this comes directly from General von Armin. This mission is to be conducted with the utmost secrecy. No one is to mention the Mothman to anyone outside of this unit. This includes writing letters to your families, even writing in your personal journals. Anyone who disobeys this order will be severely punished. Is that understood?"

"Ja, Herr Rittmeister," the men responded in unison.

"We will take off within the hour. *Jasta* Ten and *Jasta* Eleven will search for the Mothman. *Jasta* Four and *Jasta* Six will patrol for enemy fighters. Even though our priority is to find this creature, the British are attacking along the Ypres Salient. We still have to prevent them from controlling the skies."

Many of the pilots nodded.

"Any questions?"

Everyone remained silent. Manfred's lips parted, about to give out specific assignments, when Erhardt raised his hand.

"Yes, *Leutnant?*"

"Sir, I think I know what that creature is."

Every pilot turned to him.

Manfred raised an eyebrow. "What do you think it is?"

Erhardt bit his lip, gazing at the other pilots who stared back at him. His shoulders rose in a deep breath before he spoke. "It is a demon."

Voss scrunched his face. Gude also screwed up his face in doubt. Adam lowered his head and shook it, while von Döring tilted his head. Many of the other men wore similar expressions of disbelief.

"What makes you say that?" Manfred asked in a flat tone, conveying his skepticism.

"Well, um . . ." The veins protruded in Erhardt's neck. "Just look at it. The wings, the . . . evil eyes. What else can it be?"

A couple of men sniggered. Others shook their heads.

"Aren't demons supposed to be red?" asked Voss. "And don't they have horns and fangs?"

"Well, yes." Erhardt shifted in his seat. "But this could be a different sort of demon. It . . . It has to be."

"Sure it is," Voss scoffed. A few more men chuckled.

"All right, enough," Manfred ordered in an even yet authoritative tone. "The truth is we have no idea what this thing is exactly. Hopefully we will, preferably after we have killed it."

After providing his pilots with specific assignments, he dismissed them. The men filed out of the tent. Manfred followed, his eyebrows knitting together when he saw Erhardt lingering behind the other pilots. When everyone exited, he paused and turned to Manfred.

"*Herr Rittmeister.* May I please have a word?"

"Very well." Manfred nodded.

Erhardt looked away for a moment. "I know the others may not believe me, but I tell you. This Mothman, it must be a demon. Right from the pits of Hell."

Manfred studied the other pilot's face. It was stiff, his eyes narrowed. The expression radiated absolute certainty.

"I will ask again. What do you base this opinion on?"

"I . . . I . . . it has to be, sir. Please, you have to believe me."

"Why must I believe you?"

Erhardt's mouth opened, but he uttered no words.

"Well, *Leutnant?* Do you have an answer?" Manfred's tone was firm.

"I just . . . know." He visibly swallowed. "I think . . . I think we should bring in pastors, sir."

"We already have a chaplain."

"Then we should bring in more. For all we know, bullets will have no effect on a demon. Our only hope is --"

"*Leutnant,*" barked Manfred.

Erhardt straightened. *"Ja, Herr Rittmeister."*

Manfred inhaled, then spoke in a more even tone. "I know this Mothman is . . . something no one has ever seen before. Yes, it can be frightening. But we must keep our wits about us if we are to defeat this creature. And that is what the Mothman likely is. Some animal. Perhaps found by the British or French to use against us, or perhaps something that has been hidden in this part of Europe for years and only now has decided to appear. And like any animal, it is not immune to bullets. We will find it and we will kill it. Understood?"

"Ja, Herr Rittmeister." Erhardt dipped his chin, a crest-fallen expression on his face.

"Good. Now get to your plane."

Erhardt turned, grabbed hold of the tent flap, then paused. "One more question if I may, *Herr Rittmeister.*"

Manfred groaned, but said, "Very well."

"Do you believe in God?"

"Of course."

"Then if God is real, surely the Devil and his demons must be real, too."

Manfred said nothing, just watched Erhardt as he left the tent.

THIRTEEN

Manfred pulled on his gloves as he strode toward his plane. Unlike the past few days, he no longer fretted about seeing the winged monster. If anything, he hoped it appeared. Preferably, right in front of his gunsights. Then this thing of nightmares would be dead and his unit could concentrate on stopping the new Entente offensive in Flanders.

A mechanical growling caught his attention. He turned to find Voss shooting across the field on his motorcycle. A couple of times it bounced so much he thought the commander of *Jasta* 10 would flip over the handlebars.

Who needs the British or the French? The fool will probably wind up killing himself on that contraption.

He could order Voss to not ride his motorcycle. But one thing he learned over the past year was sometimes you had to indulge the eccentricities of certain pilots. Or rather, the good ones, like Voss. His recklessness on his motorcycle, his slovenly dress when not flying, went against Manfred's Prussian military upbringing. But taking those things away from him would make him unhappy, which could affect his performance in the air. Given the man's combat record, he could overlook his breaches of decorum.

Voss turned the motorcycle to the side and skidded to a stop near his D.III, the back wheel throwing up clumps of grass and dirt. One of the mechanics rolled the bike away as Manfred walked over.

"You could have walked here, you know," he said to Voss. "It's not as though it is a long trip from your tent to your airplane."

Voss shrugged. "Yes, I could have walked, but it would not be as fun."

Manfred gave a brief shake of the head, wondering not for the first time how he'd become such good friends with Voss.

Gripping the ends of his silk scarf, Voss ran his gaze over the Albatros D.III. "Who could have imagined this, *Herr Rittmeister?*"

"Imagined what?"

"That one day, instead of hunting British or French planes, we'd be hunting a flying monster. Or a demon if *Leutnant* Erhardt is correct."

"Do you believe his claim?" asked Manfred.

Voss snorted. "No. He's being ridiculous. I'm certain there is a more natural explanation to this Mothman."

Brow furrowed, Voss leaned a bit to his right, staring past him. Manfred was about to ask what was wrong when he said, "But it seems *Leutnant* Erhardt truly does believe it is a demon."

Manfred turned around. His eyebrows scrunched together when he saw Erhardt, head bowed, standing next to his plane. In front of him, one hand out, was *Hauptmann* Neuer, the unit's lanky, narrow-faced chaplain.

He walked over to the pair, Voss following. They got within a couple of meters of Erhardt when he said, "Amen."

"*Leutnant.* May I ask what you are doing?"

Erhardt jerked in surprise. "Sir. I just had the pastor bless me before this mission. Considering the . . ." He grimaced, looking away for a moment. "Given the nature of our mission, I thought it prudent."

Manfred's mouth formed a tight line. The demon talk again? Part of him wanted to quash it. He could tell at the briefing how nervous some of the pilots had been at the thought of confronting this Mothman. How fearful would they be if they believed the creature to have originated from the depths of Hell? Erhardt having the chaplain bless him before battle would serve to feed that belief.

But a chaplain praying over a pilot was not uncommon. It sometimes helped ease a man's nerves before taking to the skies. The last thing Manfred needed in the air was a scared pilot, especially for this sort of mission.

"I wouldn't mind a blessing, pastor," said Voss.

Manfred snapped his head toward the leader of *Jasta* 10, eyebrows knitted together in surprise.

Voss shrugged. "In the unlikely event I'm wrong."

Manfred said nothing as Neuer stepped over to them.

"Of course." He nodded to Voss, then looked at Manfred. "And you, *Herr Rittmeister?*"

He drew a breath, mulling over the offer. "Yes, thank you." Manfred did not believe Erhardt's claim the Mothman was a demon. But as Voss just said, what if he was wrong?

Neuer held up a hand. "Lord, please watch over these men as they embark once again into the sky to defend their brothers-in-arms and their great nation from those who seek to do them harm. Keep Manfred and Werner safe from all dangers they might encounter. Let their efforts help bring about a swift conclusion to this horrific conflict that has seized Your world. Honor and praise be given to thee, O Lord God Almighty, most dear Father of Heaven, for all thy mercies and loving kindness shown unto us. Let thy mighty hand and outstretched arm be still our defense. Amen."

"Amen," Manfred and Voss repeated.

"God be with you, gentlemen." With a brief smile and nod, Neuer ambled off.

Voss watched the chaplain depart, then turned to Erhardt. "If what you said at the briefing is correct, this is one mission where we may really need God with us."

"I believe so," replied Erhardt.

Manfred just gave a small grunt in response as Voss adjusted the silk scarf around his neck.

"*Leutnant* Voss," said Erhardt. "May I ask a question?"

"Go ahead."

"I was curious. Why do you always wear a silk scarf every time you fly?"

"Ah." Voss rubbed his thumb and index finger on one of the edges. "I simply wish to be prepared."

Erhardt screwed up his face, confused. "Prepared for what?"

"In case I'm ever shot down and captured. I want to make sure I look presentable to the girls in Paris." He tacked on a large grin.

Manfred grunted. "As much as it would disappoint the girls in Paris, please try not to get shot down and captured. You've been here little more than a week, and I have no desire to find another *Jasta* commander so soon."

Voss smiled and bowed. "As you wish, *Herr Rittmeister.*"

Shaking his head, Manfred ordered the two to get in their planes. He strode toward his Albatros D.V.

"Good hunting, *Herr Rittmeister,*" said Ewers.

"Danke, Unteroffizier," replied Manfred as he climbed into the cockpit.

The mechanic bit his lip. "Is it true, sir? Are you actually hunting a monster?"

Manfred gave a brief chuckle. "So you already know about the Mothman."

"Ja, Herr Rittmeister." Ewers grimaced. "Some of the men say it looks like it was spawned by the Devil himself."

A frown creased Manfred's lips. Apparently, Erhardt was not the only one convinced the Mothman was a demon. "I do not know if this creature is one of Lucifer's minions or if the British or French brought it back from some mysterious jungle halfway around the world to set it loose on us. But I do know one thing, *Unteroffizier.*"

"What is that, sir?"

Manfred's face stiffened. "When we do find this thing, we are going to kill it."

***

Except for the white puffy clouds, the skies over Flanders were clear. Manfred scowled. Whereas during previous flights he prayed he wouldn't see the Mothman, this time he desperately wanted it to appear. Then he or one of his men could cut it to pieces with their machine guns and they could go back to dealing with a normal, human enemy.

He swept his gaze across the sky again, and again. No sign of a dark winged monster. Though in the distance he did observe gray and black clouds hovering over the ground. No doubt thrown up by artillery. Maybe British, maybe German. Or both.

Manfred grunted. Tens of thousands, probably hundreds of thousands, of British and Commonwealth troops were throwing themselves against the German lines along the Ypres Salient. How many of his fellow soldiers had perished? How many more would die over the next several hours? The next several days? The next several weeks?

His anger boiled as he watched the mass of smoke. Could this be the Entente's plan? Release some frightful beast and cause so much fear that his superiors diverted valuable resources – meaning his squadrons – in order to hunt it down?

We may have just handed control of the skies to the British because of this damn thing.

A pulsating flame shot through his brain. Eyes shut tight, Manfred leaned forward, growling. *Not now, dammit.*

He let out quick, sharp breaths and forced his eyes open. Would these headaches ever end? Would he have no choice but to see a doctor?

And have them stick me back in a hospital bed again? Or worse, assign me to a desk?

That was not an option, not as long as the war raged. He would have to cope with it the best he could. When the fighting was finally over, though . . .

Who knows when that will be?

The pain subsided. Manfred's muscles loosened as he searched the skies again. Nothing. He checked the other planes nearby. No signal from anyone that they had spotted the Mothman.

They kept flying. Shell craters and snaking trenchlines became more prominent below. How much longer before they passed over the British lines? Could they be over them now? Even if they were, they were high enough to be out of the effective range of enemy groundfire, something he always drilled into his pups.

Another search of the skies turned up nothing. Manfred sighed. He had a feeling the Mothman would not reveal itself today. *Hopefully tomorrow.*

What if it didn't? What if they searched for a week and still couldn't find it? How much would that help the enemy with its assault on the –

Movement to his left caught his attention. Gude's plane waggled its wings. He jabbed a finger to the southeast.

Manfred leaned forward in his cockpit. His chest tightened. Four dark shapes appeared on the horizon.

Four of them? One had been bad enough. Now they had to deal with four?

He pushed aside his shock and angled the D.V's nose toward the four dots and accelerated. Erhardt, Gude, and the remaining three aircraft followed. His eyes bore in on his targets. Now he didn't care how many of these creatures roamed the skies. He and his men would finish them.

Manfred gripped the control stick tighter. The objects in front of him took on more distinct forms. They were not Mothmen, but other aircraft. Disappointment flared through him. He stamped it out quickly.

They may not be monsters, but they were planes. British Bristol F.2s.

And they were turning toward Manfred's group.

FOURTEEN

Manfred scowled. That was one of the major problems with this mission. Along with trying to find a single, man-sized flying creature somewhere within the one million hectares of Flanders, there was still a war going on. The British were not about to leave them alone while they searched for the Mothman.

He jammed the stick left. The other five planes followed. The turn lasted mere seconds. Manfred yanked the stick right, hoping to come around behind the Bristols.

But the British had matched his turn. He grimaced. Though larger than the D.Vs and DIIIs, the Bristols proved incredibly maneuverable. More so than Manfred's own plane. But the D.V held a slight edge in speed and rate of climb, and he knew how to use them to his advantage.

He pulled back on the stick. The engine screamed. Wind howled over him. An invisible anvil pressed down on his chest. Just before he thought it would crush him, Manfred nosed the D.V over.

A Bristol passed below. Manfred triggered a quick burst. He sneered when no smoke or flame gushed from the enemy plane.

Manfred pulled out of the dive. He clenched his teeth against the mounting pressure on his body, standing the D.V on its right wing. The Bristol swung to its left, the pilot probably hoping to catch Manfred coming out of his dive.

It won't be that easy, my friend. He heaved the plane onto its left wing and turned toward the British fighter. Flashes spat from the observer's machine gun. None of the rounds hit him. The observer had been too far away for a good shot. Probably new to combat. Manfred hoped the pilot was just as much a novice.

Manfred climbed again. He glimpsed the other German and British airplanes twisting through the sky, orange spouts erupting from their machine guns. He checked below. The Bristol had its nose up, rising to meet Manfred.

But he had the altitude advantage. He pushed the D.V over into a dive. The gunsights settled above the Bristol. Leading it . . . leading it . . .

Manfred fired. Splinters of wood leapt off the rear of the enemy fighter.

The D.V shot past the Bristol. Manfred twisted in the cockpit. The Bristol turned away, trying to escape.

He smiled. The British pilot just violated one of the main tenets of *Dicta Boelcke*. Never run away from the fight. Always turn into your attacker and meet him head-on.

Manfred swung around and accelerated. He soon got on the Bristol's tail, then dropped a bit below him. He was not about to give the rear gunner a good shot.

The British fighter swung right. Manfred matched the turn. The gunsights slid over the Bristol's tail. Manfred accelerated, shoulders tensed. He was about to get his first kill since suffering that damned head wound. Just a little closer and –

A flicker of movement caught his attention. He glanced right and held his breath. Something dark glided across the sky. Something with wings.

The Mothman.

Manfred stared back at the Bristol. He almost had the other plane. To break off now would go against everything he preached about aerial combat.

But General von Armin had made killing the Mothman his unit's priority.

An enemy plane in front of him. A mysterious creature that had proven deadly to both German soldiers and airmen.

Scowling, Manfred jerked the D.V right and dove on the winged form. He leaned forward in his seat, eyes narrowed as he put the nose on his target.

Which flipped over, then swayed to the left.

Manfred's brow wrinkled. That's not how a flying monster should be able to maneuver . . . or how he imagined it could maneuver.

The object flipped over again. That's when he recognized it. It was not the Mothman. It was a piece of wing, apparently shot off from one of the other planes.

He slammed his back into the cockpit, seething and spitting out a curse. How could he have been fooled by a piece of debris? Fooled so badly that he abandoned his attack on an enemy fighter . . .

Manfred's spine went rigid. The Bristol!

He swung his head around. The British fighter dove on him. Orange flashes spewed from its forward-mounted Vickers machine gun.

Two jolts shook Manfred's D.V. He yanked back on the stick. The plane shot into the air. Teeth bared, the wind blasted over him. His body grew heavier by the second.

Manfred rolled his plane and dove. The Bristol roared right at him. No time to aim properly. He triggered the machine guns.

The British fighter flashed beneath him, a handful of meters separating them. He let out a quick sigh of relief at just missing his opponent. He did not have time for any other reaction. Not with the other pilot after him.

I should have shot him down before breaking off. Stupid!
Chastise yourself later, Manfred.

He snapped his head left to right, checking the sky behind him. He gritted his teeth when he didn't see the Bristol.

Manfred pulled out of the dive and swung left. His eyes darted around the sky. Where was the damned fighter?

A look behind. Nothing. A glance up.

The Bristol screamed down on him.

Manfred pulled up the D.V's nose. He checked behind him again.

The enemy plane climbed after Manfred. His body stiffened, waiting for the torrent of bullets. *Is this the end?*

Splinters leaped off the front of the Bristol. Smoke and flame burst from the engine. The fighter spun and plummeted toward the ground.

A black and orange Albatros DIII soared above him. Erhardt's plane. The *leutnant* swung back toward Manfred and waggled his wings. He returned the gesture.

Manfred searched for another target, but two of the Bristols were flying back to the Entente lines. He saw no sign of the remaining British fighter. One of his pups must have shot it down.

With the fight having consumed a lot of fuel, he had the other planes form up on him for the return home. Jaw set, he stared straight ahead, stewing over missing out on another kill.

All because of a stupid mistake. A mistake that almost cost him his life.

The throbbing in his head prevented any more self-recrimination. The stabbing pain in his skull lasted the rest of the flight back to their base. When Manfred landed and got out of his plane, he kept his face a tight mask, not wanting to betray his discomfort, as Erhardt strode up to him.

"It appears I owe you my life, *Leutnant.* I hope now you realize the importance of working in tandem."

"Ja, Herr Rittmeister." Erhardt nodded, then tacked on a grin. "So will this get my name in the newspapers back home? 'The Man Who Saved *Der Rote Kampfflieger.'"*

Manfred gave a slight chuckle. "Who knows? That might make you more famous than me."

"Somehow I doubt that."

With a brief smile, Manfred turned and started back to his tent.

"Herr Rittmeister," said Erhardt. "Permission to ask a question?"

"Granted." Manfred nodded.

Erhardt looked around, as though making sure no one was in earshot. He then took a few steps closer to Manfred, lips tight for a moment. "I was curious, sir. I saw you behind that Bristol. You had a clear shot at him, then broke off for some reason. You have stressed finishing an attack once you have started one, but you didn't. I'm curious as to why."

Manfred glanced at his boots for a second. Part of him wanted to tell Erhardt to drop the matter. A commander admitting failure to his men could be devastating to morale. People like him were supposed to make the correct decisions, always. He had not during his fight with the Bristol.

But Erhardt had seen what happened. Dismissing the question could cause him to question his judgment. And soldiers talk, which might cause more pilots to have concerns about their commanding officer.

In this case, admitting a mistake might be better than hiding it.

"I saw what I thought was the Mothman. Given General von Armin's orders, I felt that was the priority and went after it. It turned out it was debris from one of the Bristols we shot down. Even if it was the Mothman, I should have known better. At that moment, the British plane was the more immediate concern. I should have shot it down before going after the Mothman. Rather, what I thought was the Mothman."

"Ah." Erhardt stared to the side for a second. "I think you made the right decision."

"No I did not," Manfred said in a firm tone. "As you saw, that Bristol almost shot me down. We cannot ignore one threat to go off and pursue another."

"But the general's orders . . ."

"Yes, we have our orders, which I would have obeyed . . . after I had dealt with the Bristol."

"Yes, *Herr Rittmeister.* I understand. But this Mothman. We have got to kill it."

Manfred noted the zeal in Erhardt's voice. He figured it had to do with the man's belief the thing was actually a demon.

"We will kill it, *Leutnant.* But we have to be smart about it. Do not treat this as some holy crusade. That means if you have an enemy plane in your sights and the Mothman appears, do not chase after it until you have dealt with the threat in front of you. I do not want you making the same mistake I did, because if you do, you may not be as fortunate as I was to learn from it."

FIFTEEN

"Three days we've been looking. I wonder if we'll ever find it."

Voss's words prompted Manfred to look up from his meager breakfast of bread and eggs. "We will. Because we have to."

From across the table, the younger pilot snorted. "And the British could overrun our lines before we find this monster. We are flying in circles day after day doing nothing while they rule the skies. This is a waste of our skills." Voss tacked on a scowl.

"Perhaps you should send a dispatch to General von Armin." Von Döring looked up from his coffee cup, an edge to his voice. "I am sure the complaints of a lowly *leutnant* will get him to change his mind." He lifted the cup toward his mouth, then stopped. "Better that he deal with your constant whining than me."

"Are you happy letting the British fly over the battlefield unmolested?" Voss' hands snapped up. "Shooting down our planes while the Army's best pilots fly around aimlessly."

"It is not aimless flying." von Döring glared at him. "We have been given a mission, and we are carrying out that mission."

"And how long will this mission go on? Another week? A month? Or until we are unable to search for this Mothman because British planes are swarming the sky because we haven't been shooting them down like we are supposed to."

Other pilots in the mess tent glanced over at Voss and von Döring.

"It will go on for however long it must, and all your complaining will not change that."

"So when --"

"Enough." Manfred's sharp tone cut him off. He gazed at the other tables. The pilots who'd been staring at them went back to eating their breakfasts.

He leaned over the table toward Voss and von Döring and spoke in a low voice. "I will not have two of my *Jasta* leaders arguing in front of the men. Have you any idea how bad that is for morale?"

Neither man answered, just lowered their heads.

"This mission will continue until we kill the Mothman or until General von Armin says otherwise. Understood?"

"*Ja, Herr Rittmeister,*" they both replied.

"No more complaining. No more arguing," said Manfred. "We need to show everyone in the *Jagdgeschwader* that we are united in purpose and we will succeed. Understood?"

"*Ja, Herr Rittmeister.*"

Satisfied, Manfred leaned back and bit into a piece of bread. His gaze shifted between the two men across from him while he chewed. Voss' complaints did not surprise him. Inaction did not suit the young pilot. The fact that he had never learned when he should keep his mouth shut did not help matters.

Von Döring, though, was more level-headed. So much so that Manfred entrusted him to take over the unit during his convalescence. To be baited into an argument like that showed how frustrating this situation had become.

Manfred studied the other pilots. If Voss and von Döring were this upset, he could only imagine how the rest of *Jagdgeschwader 1* felt. He could order them to stop complaining as much as he wanted, but once they were out of his sight and in their tents, they would voice their displeasures. That's what soldiers did when out of earshot of their commander.

Even he could not deny his own frustrations. British attacks had increased along the Ypres Salient. Both sides still vied for control of the air. Each day that passed without finding the Mothman meant more opportunities missed to thin the ranks of the Royal Flying Corps. For him personally, he had not had one kill since returning to the unit. The dry spell did not sit well with him.

Manfred's fork hovered above his eggs, mulling over Voss' complaints. His friend had some valid points. His pilots were better suited for aerial battle. A dedicated scouting squadron should be searching for the Mothman.

But even with his fame, he doubted he would be able to convince General von Armin of that.

***

Manfred lifted his head as he walked to his plane. Several thick clouds stretched across the sky. White with some gray mixed in. The meteorologists believed rain would come soon. Maybe later today, most likely tomorrow. Combined with the rain that fell last week, it would make conditions in the trenches miserable.

Or rather, more miserable than they already are.

He spotted *Hauptmann* Neuer giving a blessing to Erhardt while standing by his DIII. It had become routine for Manfred's wingman before every mission. Not only Erhardt, but several other pilots. Manfred himself had not taken part in that, though he had prayed a couple of times for the Lord to help them find the Mothman. Hopefully, that prayer would be answered today.

He climbed into his plane as one of the C.III reconnaissance planes on loan to the unit rolled down the field. Their cameras had photographed numerous trenches and rubble-strewn villages within the region of known Mothman attacks. So far they found no evidence of the creature's nest.

The D.V's engine sputtered, then buzzed to life. Soon he was airborne, the other planes of *Jagdgeschwader 1* following. Manfred would take his four-plane formation and patrol the area around the Gheluvelt Plateau east of Ypres. A turn of the head showed Erhardt flying off his left wing. Just beyond and above flew Heldmann and his wingman, Hans Wolff, one of the newer pilots in the unit. He aimed the nose of his fighter south, his eyes already searching the sky, hoping to spot the telltale black form and moth-shaped wings.

He did not.

The planes neared the Ypres Salient. A brown-gray ribbon stained the horizon. Manfred figured it smoke and dust thrown up by artillery barrages. Someone had launched a major assault in this part of Flanders.

He shuddered, remembering his own experiences with shells exploding around him in Russia and France. The deafening, never-ending crashes. The helplessness. Unable to shoot back. Just curled up in a hole and praying a shell did not land on him.

His sympathy swelled for the men going through that hell down there. Closing his eyes, he rid himself of the feeling. He could not afford to dwell on it. He and his men had a job to do up here, as the soldiers in the trenches had their job.

He lifted his gaze from the distant smoke. Except for the thick clouds, the sky was empty.

The four fighters continued on. Every so often Manfred glanced at the battlefield below. The clouds made the ceiling lower than he would have liked, but they should still be out of the range of effective ground fire.

Still, no sense in taking chances. He turned his D.V right to skirt the frontlines. A look behind showed Erhardt following.

Heldmann waggled his wings. Manfred just made out the pilot's arm extended toward the southwest.

He swung his plane around. Squinting, he peered over the propeller. Where . . .

There! Four shapes moving in the distance. Aircraft. Coming from the south, they had to be Entente.

Manfred shimmied his stick left to right, rocking his wings. He pulled up, leveling out just under the cloud bank. He checked the other planes. Erhardt, Heldmann, and Wolff were with him. The enemy aircraft continued flying straight, apparently unaware of their presence.

He smiled. The element of surprise was on their side. Maybe now he could end his dry spell.

Manfred eased the Albatros D.V around, getting above and behind the enemy formation. Stubby biplanes. Probably DH.5s. Not the best fighters in the British inventory.

His smile grew wider.

Manfred dove. So did his three pups. The engine screamed. The wind blasted over him. He lined up his guns on one of the trailing planes, still flying straight. It grew larger with each passing second.

He thumbed the firing buttons. The forward machine guns chattered. Splinters jumped off the side of the DH.5. It wobbled as Manfred's plane shot past it.

He pulled out of the dive and swung around, the pressure nearly crushing his body. He clenched his gut and groaned as he eyed the scene before him. The DH.5s scattered as his men dove past them. One of the enemy planes trailed smoke, but remained airborne. The one he had shot. It turned back toward the Entente lines. Manfred gripped his stick tighter. The hunter in him screamed to pursue the crippled plane and finish it off. All his dictates about target fixation and not pursuing aircraft over enemy lines pushed back.

It was those concerns that won out. Much as he wanted a kill, the DH.5 was out of the fight. Chasing it over British lines would be too risky.

He swung his D.V back to the fight. He glimpsed Erhardt's D.III being chased by a DH.5s. But Erhardt made a sharp turn left, then climbed. His pursuer had a hard time matching the maneuver. Manfred nodded. His wingman would soon be on the enemy's tail.

Manfred checked his right. Another DH.5 raced across the sky, heading in Erhardt's direction. Manfred turned to engage.

The DH.5 banked right. It must have seen him.

Good. Not only did he keep him away from Erhardt, the enemy turned his tail toward him, making him an easier target. Manfred increased his speed and closed on the fleeing plane.

Something caught his eye. A disturbance in the clouds. He glanced up . . . and the breath stuck in his throat.

A thick black form appeared, its huge wings spread out.

The Mothman . . . and it was diving right at him.

SIXTEEN

Manfred shoved the control stick left. The D.V stood on one wing. Manfred gritted his teeth, not letting up on the sharp turn. Invisible forces crushed his body.

The plane arced around. It took great effort to turn his head left to right, searching for the Mothman. Manfred had hoped the tight turn would put him on the creature's rear.

But he could not spot it.

He leveled out, the invisible pressure abating. Manfred searched above him. Nothing. A check to the left, then the right, and the rear. Still no sign of the Mothman.

"Dammit." He sneered. He couldn't have lost him so quickly. Where could it be? Maybe it had gone back into the clouds, looking for another chance to dive on him. But would it break off an attack so soon?

Not above, not behind, not to the sides. So where . . .

Manfred's eyes widened when he realized it. He looped the D.V over, flying upside-down.

The Mothman flew straight up at him.

Manfred rolled the plane. He'd just leveled out when something struck the wing. The D.V shook and spun over. He gripped the stick, fighting to move it to the center. The world whipped around him. He lost all sense of direction. Panic enveloped him. Tremors raced up and down his body. This was it. He would die here. A scream grew in his throat, begging for release. Images of his parents and his brother Lothar flashed across his mind. He'd never see them again.

Pain sliced through his brain. The paralyzing terror dissolved. Training and instinct returned. He worked the stick until the plane straightened out.

That's when the ground rushed up toward him.

Manfred pulled the stick back. The engine screamed as the nose rose. The plane shook. All his muscles tightened, expecting the aircraft to break apart.

He flew, higher and higher, then leveled out. His shoulders loosened, and he let out a breath . . . then scolded himself. Now was not the time to relax.

Manfred looked right. The tip of his lower wing had been shredded. It had to be the Mothman's doing.

Where . . .

He gazed up, then behind. His heart thudded.

The monster was on his tail and gaining.

Manfred banked right, hoping to swing around and get the Mothman from behind. But the D.V felt sluggish. To be expected with a piece of his wing damaged.

He checked behind him. The Mothman no longer pursued him. He gazed left to right as he completed the loop. Again, no sign of the thing. Manfred craned his neck.

The monster dove at him.

He banked right. The Mothman streaked past him . . . then looped over, rolled, and darted after him.

Manfred's eyes bulged. He had never seen anything able to maneuver like that. And the speed. He doubted any bird could fly so fast. Fast enough to match his Albatros D.V. His eyes fixed on the monster as it soared past his tail.

That's when its eyes flashed blood red.

Manfred struggled to take a breath, fought to look away. He couldn't. Those evil eyes held him, reached into his soul. All thought vanished, replaced by horror. The horror of this thing, this demon, slashing his chest open. Blood spilling over him. Nothing he could do except . . .

A knife of pain cut through his skull. He closed his eyes and turned away. Manfred grunted, trying to fight through his fear.

A jolt battered his D.V. Manfred snapped his head left. The Mothman was alongside him. It must have slashed the side of his plane, like it had the C.III he and Erhardt had found.

He snapped the D.V right. Manfred sneered as he increased his speed. Running from an opponent violated one of his most important tenets of aerial combat. But trying to turn into the Mothman did not seem an option. Not with its incredible maneuverability. He had to come up with another tactic.

Manfred checked over his shoulder. The creature flapped its large wings, propelling it closer to him.

He banked left, then right. The Mothman kept pace. Not only that, it got closer. The veins in his neck stuck out. What if it attacked his rudder? He'd have no way to control his plane.

Another glance over his shoulder. The Mothman was just a few meters from his tail.

Manfred looked ahead. He couldn't outmaneuver the thing. It was almost as fast as his D.V, maybe even faster. What else could he do?

He looked back again. The Mothman drew closer. Another few seconds and it could swat his tail.

Manfred swung his head forward. His gaze settled on the smoke and dust drifting over the battlefield below. Here and there he saw dark spouts shooting up from the ground. Shellbursts.

His jaw stiffened. It would violate another of his tenets, but right now he had no other options.

He pushed the nose down. The D.V screamed toward the ground. Manfred glanced behind him. The Mothman followed.

Good. Flying this low to the battlefield meant a greater chance of being hit by ground fire. But it also meant the Mothman would share that risk.

Lothar would be proud. His brother had always been the more reckless of the two. But right now, such recklessness might be the only thing to save him.

The D.V roared closer to the earth. Part of the ground below exploded. His chest tightened, imagining a random shell shattering his plane. But he was committed to this insane scheme.

He pulled up less than one hundred meters from the ground. He'd just leveled out when he swung the plane left. Flying straight only made him an easy target for British soldiers. He looked behind him . . . and jerked in surprise.

The Mothman frantically flapped its wings, flying in place. It clutched its head, twisting from side to side. Had someone on the ground shot it?

The creature turned away from the frontlines, swaying as it flew. Manfred imagined this would be what a drunkard with wings would look like.

It also meant the Mothman was vulnerable.

He swung the D.V around and brought the nose to bear on the wobbly monster. Manfred tried to anticipate its movement as he closed with his target.

Wait . . . wait . . . now!

The machine guns rattled. The Mothman twitched, its arms splayed out. It then crumpled and dropped from the sky.

Manfred jerked straight up in his seat, elation shooting through his body. Eyes wide, he sucked down an energetic breath and watched the beast plummet to earth.

Dead. Finally, it's dead!

Manfred circled above the beast until he lost sight of it amidst the ground clutter. He took note of any landmarks to aid the search parties. General von Armin would certainly want to recover the body.

He gained altitude as fast as possible, then leveled out and leaned back in his seat. He let out a long exhale. For days this seemed an impossible task. Find one creature in the enormity of the Flanders Theater.

But he had done it, and he had killed the damn thing.

Kill number fifty-eight. Manfred chuckled, thinking about the silver cups back at his tent that commemorated his victories. What would the maker think when he engraved on the next one *Mothman?*

SEVENTEEN

"I have killed the Mothman."

Seconds after Manfred made the proclamation, the other pilots and mechanics cheered and lifted him on their shoulders. The jostling sent the world around him spinning. His skull pounded. But he gritted his teeth, trying to endure it. This was cause for celebration.

They soon set him down. Manfred's legs trembled. His hand clamped down on the shoulder of the man nearest him, Hans Adam. Manfred hoped the other pilot took it as a friendly gesture and not his commanding officer doing all he could to stay on his feet.

His face scrunched in a mix of anger and embarrassment. When was this going to end? When could he finally fly without these damn headaches? Without feeling like he would collapse whenever he climbed out of the cockpit?

The continued cheering of his pups chiseled away those negative thoughts until they vanished.

"Well done, sir." Erhardt practically jumped in his way, beaming. "You've truly done the Lord's work."

"Danke, Leutnant." Manfred returned the smile.

Werner Voss slapped Erhardt on the back. "Still think that thing was a demon, eh, Fritz?"

"What else could it be?"

"And you think mere bullets could have killed something that rose from the depths of Hell? Even bullets from the guns of *Der Rote Kampfflieger?"*

"Apparently." Erhardt glared at Voss. "And are you suddenly an expert on demons?"

"Are you?" Voss belted out a laugh.

Erhardt grunted and turned away. Manfred heard the pilot mutter, "More than most."

Manfred raised an eyebrow. What had he meant by that?

"If you'll excuse me," he said to the two pilots. "There are reports to be done on this engagement."

Voss scoffed. "Yes, there are always reports commanders must do, on everything."

"As though you would know, considering how infrequently you submit your reports. That needs to change, Werner."

Shoulders sagging, Voss sighed. *"Ja, Herr Rittmeister."*

Manfred eyed his friend for a few seconds, wondering if he would actually do it. Since coming to the wing, Voss had gone out of his way to avoid the more mundane aspects of command. He could not indulge him in that much longer.

Manfred headed for his tent, where Moritz stood by the entrance wagging his tail. He patted the dog's head and entered the tent. The headache returned, squeezing his skull. He grimaced, eyeing his desk. He needed to write this report immediately. But the throbbing in his head, the sudden heaviness in his muscles . . .

He stared at the desk, then at his cot. He frowned and shook his head.

Your bad habits are wearing off on me, Werner.

He trudged over to the cot and lay down. Within minutes, he was asleep.

***

Manfred awoke two hours later, his headache and fatigue gone. Rubbing his eyes, he splashed water on his face from his canteen, and sat at his desk. He wrote his report as quickly as possible, including his best estimate on where the body of the Mothman landed, and summoned a messenger to take it to General von Armin. As he watched the young soldier ride off on his motorcycle, Manfred shook his head, scolding himself. He should have written that report the moment he returned to his tent. What sort of leader sleeps instead of informing his superiors that a creature that had killed or driven mad several soldiers was now dead? Why the hell did his body keep betraying him?

And what can I do about it? Going to Dr. Kretschmer was out of the question. He would recommend to the generals that he be grounded. Manfred could not afford to be out of the war with the British attacking the Ypres Salient. All he could do was pray these ailments went away on their own.

And what if they do not? Manfred tightened his body, holding back a shudder. Fear of never flying again clutched his mind. He exhaled loudly a couple of times, trying to ease his worries. At least his fear of not flying was nowhere near what he had felt when the Mothman had flown next to his plane.

Why had he been terrified so? Yes, being shot at could scare any man. Manfred was no exception. But he could not remember the feeling being as intense as it had been when the creature attacked him.

Perhaps it has something to do with the headaches.

Manfred snorted. Enough of these wonderings. He still had other reports to attend to. Whatever ailed him, he would be damned if it kept him from his duties.

***

The pathetic offerings for dinner – thin sausages, bland potatoes, and a slice of bread – did nothing to dampen the mood of the men in the mess tent. The wine provided by Voss also helped. Where and how he got it, Manfred did not know. Nor did he care. Everyone enjoyed themselves as they drank and listened to him recount how he ended the Mothman's reign of terror.

"So you actually flew close to the ground?" said Heldmann. "After telling us countless times we should never do that?"

Manfred gave a slight shrug. "I still would not recommend it. But at the time, I saw no other option."

"And it worked," blared Voss. "That's all that matters."

"It worked *this* time," replied Manfred. "Someone on the ground must have hit it. The thing grabbed its head and started twisting from side to side."

He pressed his hands against his ears and turned from one side to the other to imitate the stricken monster. "When it tried to fly away, it looked unsteady. That is when I came in from behind and shot it." He used his hands, with his thumbs and pinky fingers extended to re-enact the intercept.

"Ha!" Voss banged his hand on the table. "Not only do the British and French need to fear *Der Rote Kampfflieger*, but so do monsters. No one in the sky is safe from him."

The pilots let out cheers and laughs. Farther down the table, Manfred noticed Dr. Kretschmer observing him. The staff surgeon sat unsmiling, his expression dour. Or thoughtful? Manfred could not tell.

He ignored the doctor, continued with his story, and finished his dinner. He returned to his tent to catch up on his paperwork by candlelight. Gude's violin provided background music. Mendelssohn's "Bartholdy Symphony Number Four." A joyous, inspirational piece. Appropriate for this triumphant day of flying and fighting.

Manfred was in the middle of reading a requisition form when Kretschmer appeared at the tent entrance.

"Are you busy, *Herr Freiherr?*" The doctor frowned briefly. "Silly question. A man in your position is always busy."

"I think I can spare a few minutes for our staff surgeon." Manfred laid down the paper and waved Kretschmer to the seat in front of the desk.

Taking a drag on his cigarette, he sat down.

"What can I help you with, Doctor?"

"I was listening to your story about how you killed the Mothman and . . . well, there were some aspects of it that made me curious."

"Such as?" asked Manfred.

Kretschmer stared at his cigarette as he rolled it between his fingers. "You said the Mothman was wounded by groundfire."

"That is what I assume."

"You also said the creature grabbed the sides of its head, correct?"

"Yes." Manfred raised an eyebrow, wondering where the doctor was going with this.

"Mm." Kretschmer gave a slow nod. "I find it curious."

"What?"

"If the Mothman was hit in the head, it would have likely died. Monster though it is, its build is similar to a human's. Which would indicate the same basic physiology. Heart and lungs in the chest, stomach in the mid-section, and brain in the head."

"Maybe the bullet grazed its head," Manfred offered.

"One grazing wound I could accept. But two?" Kretschmer shook his head. "You went like this." He clamped his hands over both ears. "Which to me would indicate two wounds, on opposite sides of the head. The chances of both the Mothman's . . . ears, I guess, being hit at the same instant . . . I cannot even calculate that."

"Maybe it was hit somewhere else."

Kretschmer's brow wrinkled in a doubtful look. "You have accidentally bumped your knee at some point in your life?"

"Of course," replied Manfred.

"Where does your hand go when you do that?"

"To my knee."

"And when you get a cut on your finger, what do you do?"

"I press down on it until I can wrap it in something."

"Exactly." Kretschmer stuck out a finger for emphasis. "The first instinct is to touch the area of a wound, whether it is a human or an animal. So why would the Mothman grab his head if he had been shot elsewhere?"

Manfred stared at the doctor, biting down on his lip, trying to come up with an answer. He failed.

"So what do you think the reason for that is, Doctor?"

Kretschmer shrugged. "I am in the dark as much as you appear to be, *Herr Freiherr*. What might help is if I could examine the creature's body once it is found."

"I imagine General von Armin has his own doctors and scientists who will study it," said Manfred.

"Perhaps. But the general did give this unit the task of hunting down the Mothman. And as unit staff surgeon, I think any examination of the creature would fall under my purview."

"Do you think General von Armin will accept that argument?"

"He might if you present it to him. Your word does carry a lot of weight these days."

Manfred briefly grinned. "First thing tomorrow, I will dispatch a messenger to the general personally requesting that you examine the body of the Mothman upon its recovery."

Kretschmer responded with an even bigger smile. *"Danke, Herr Frieherr.* I can only imagine how this could change what we know about the an--"

A rifle shot rang out.

EIGHTEEN

Pistol out, Manfred dashed out of the tent. Kretschmer followed, with Moritz close behind.

Two more shots cut across the air. Then two more.

"What the hell are they shooting at?" blurted Kretschmer.

"We'll find out soon enough."

Manfred pounded across the airfield. He glimpsed other silhouettes rushing about.

Another crack echoed in the night. He spotted an orange muzzle flash near the perimeter. Two more blossomed in the dark. Manfred noted the angle of the shots and furrowed his brow. They had not been low to the ground as if engaging advancing soldiers. The shots had been high, aimed at the sky. *Air raid* went through his mind, but airplanes rarely flew at night.

It couldn't be . . .

A rifle cracked in front of him, the flash illuminating the young soldier holding it. Manfred recognized him. Private Uhlig. And he was shooting at the sky.

"Private!" Manfred shouted. "What's going on? What are you shooting at?"

The lanky soldier swung around, the whites of his wide, fearful eyes standing out in the darkness. "Something's flying over us. I think . . ." His voice quivered. "I think it's --"

"Up there!" Someone hollered from behind.

Manfred whipped around. Voss pointed to the sky, flanked by Gude and Erhardt. Manfred lifted his head. All his muscles froze.

No. Impossible.

Two blood-red orbs soared above them.

No. I killed you, you bastard.

The red eyes wheeled back toward them.

Manfred brought up his Luger and fired. Uhlig also fired. The red eyes veered to the right and vanished.

"Is it the damn Mothman?" Voss hurried over.

"How?" Gude threw an arm out at his side, looking at Manfred. "You killed it."

"Maybe not." Erhardt jerked his head left and right, frantically searching for the eyes. "It is a demon. How can bullets hurt a demon?"

"Enough of that talk, *Leutnant,*" snapped Manfred. "Watch the skies. Be ready when it appears again."

Clutching his Luger, Manfred's gaze roamed the night. He saw nothing. The Mothman's skin was the perfect camouflage. Panic bubbled in the pit of his stomach and slithered through his veins. The hair on the back of his neck stood. What if the damn monster dropped from the sky and grabbed him? Flew him off to its nest, where it would . . .

Cold pricks raced up and down his back.

"Shit!"

Manfred spun around. Uhlig was crouched on the ground, his hand frantically patting the grass.

"What's wrong?"

"I dropped my clip." The private's voice cracked. "Where is it? Where is it?"

"Calm yourself," demanded Manfred. "You are no good to us if you panic."

Uhlig stared up at him. He drew a couple of deep, quick breaths and returned to trying to find his Gewehr's five-round stripper clip.

More gunfire erupted farther down the perimeter. Several muzzle flashes lit up the darkness. High above, two small red circles floated over the base. Could it be looking for some weakness in their defenses before it attacked? Why was it here in the first place? He had shredded the thing with his machine guns. Nothing could have survived that.

Nothing that we know of.

"I found it." Uhlig stood, the clip pressed between his fingers. "I found it."

He went to push the clip into the Gewehr's magazine.

Something whooshed to Manfred's left. He thought it was the wind, but it sounded . . . different.

He turned. A dark mass rushed out of the darkness. The whooshing grew louder.

Manfred's stomach clenched. Cold wrapped around his spine. He opened his mouth, trying to shout. His throat seized up.

The dark mass streaked behind Private Uhlig. Something sprayed out the back of his neck. The young man went rigid, his mouth and eyes wide open.

Then he toppled forward.

Kretschmer dashed over and examined Uhlig. The doctor looked up at Manfred and shook his head.

Manfred's eyes settled on the dead private. His legs shook as he watched blood flowing from the gaping wound in his neck.

That might be me next. There was no might. It would be. The Mothman would come back and tear open his neck. Nothing he could do but wait for death to claim him.

No. No. That couldn't happen. He couldn't . . .

"It's coming back!" Gude fired his pistol until it clicked empty.

"Everyone down!" Manfred threw himself on his stomach. Voss, Erhardt, Gude, and Kretschmer did the same.

A strong wind washed over Manfred. He barely suppressed a whimper. He shut his eyes, not wanting to roll over. The Mothman could be right over him, its red eyes burning into his soul. He just wanted to lie here, face down, not confront this horror.

Get up. A voice whispered in the back of his mind.

Damn you, get up. You're a soldier.

Manfred rolled over, gritting his teeth. All the times he'd been in combat, he could not remember being this scared.

You won't survive by being a frightened little girl.

He pushed himself to his feet and searched the darkness. No sign of red eyes.

"Flares!" he hollered. "Someone send up flares!"

Seconds passed. No flares blazed in the sky. Rifle shots cracked on the other side of the base.

He looked down at the others. "Up! Get up! Time to fight this thing."

Voss, Gude, Erhardt, and Kretschmer all stood. Erhardt visibly trembled.

"We're not going to kill that monster with these things." Manfred held up his Luger. "We need to get to a machine gun."

"I'll get the private's rifle." Voss frowned. "The poor man won't need it anymore."

Manfred nodded and looked at Erhardt. The pilot stared into the darkness, still shaking.

"Erhardt. Erhardt!" He stomped over to him and grabbed his shoulders. "Get a hold of yourself, man!"

Erhardt exhaled panicked breaths. "It . . . It's going to kill us. Kill us and take our souls to Hell." His voice broke. Erhardt appeared seconds away from crying. "It will torture us for all eternity. It will --"

Manfred slapped him across the face. "Dammit, you're a soldier! An officer of the Imperial German Army. Now act like one and do your duty!"

Erhardt's jaw quivered. He swallowed hard and nodded.

The men ran across the base, Manfred leading the way. Still no one had launched a flare. But Manfred had the layout of the base memorized and knew where all the machine gun emplacements had been set up.

The outline of an arc of sandbags appeared in the darkness. They hurried into it, Manfred taking position behind the MG 08. He cocked the handle twice and peered into the darkness. No red eyes or silhouettes were visible. But the Mothman was out there, somewhere. It had to be.

"Flares!" He roared again. "Flares!"

Several seconds passed. The darkness remained. How could no one have sent up a flare by –

Two bright red orbs pierced the darkness in front of him.

"There!" Erhardt screamed. "There!"

Manfred stared at the eyes. His thumbs hovered above the machine gun's firing button . . . and stayed there. Every muscle remained frozen as he stared straight ahead. He thought about the crew of the C.III, of Private Uhlig. Blood spilling from their bodies. Would that be his fate? No, no it couldn't be.

Fire. His mind railed. *Fire!*

Fear kept him paralyzed.

The red eyes rose slightly. The Mothman leaned its head back. Then the red orbs shrank in size, as if the creature was lowering its eyelids.

Cold sweat flowed over Manfred's body. What was this thing? Could it be a demon like Erhardt said? What chance did he have? A sob gathered in his throat. He was going to die. Die horribly.

No. Please, Lord. No.

The eyes moved forward.

Deep barks burst from the darkness. The eyes swept to the right.

Manfred's terror dissolved. He swung his head toward the barking. A large, four-legged shadow stood nearby. Moritz.

He turned back to the Mothman. The red eyes returned their focus to him. In a flash, they charged.

Someone cried out behind him. Voss. He stood just above Manfred and swung his rifle. The butt thudded against the Mothman's head. The large silhouette stumbled to the side and dropped to a knee.

Manfred swung the tubular machine gun. The dark mass that was the Mothman could not be more than two meters in front of him.

The MG 08 chattered. The Mothman shuddered, then crumpled onto its back.

Erhardt rushed out of the machine gun nest and stood over the dead creature. His shoulders jerked up and down. Nervous breaths, Manfred guessed. He certainly couldn't hear them over the ringing in his ears from the MG fire.

Erhardt pointed his Luger down and fired until the eight-round magazine ran dry. Manfred thought about scolding him for a waste of ammunition, but when it came to something like the Mothman, some overkill might not be a bad idea.

A white, miniature sun formed over the base. Everyone looked up at the flare.

"About damn time," muttered Voss.

"There!" Gude stabbed a finger at the sky.

Manfred held his breath, eyes unblinking. Disbelief hit him like a fist.

Another Mothman soared over the base perimeter.

"There's more than one of these damn things?" Voss gawked at the creature.

Manfred did not answer. His astonished gaze remained on the other Mothman. So the one he killed today did not magically come back to life. It was just one of . . . *How many?* He swallowed.

Soldiers shouted and fired rifles and pistols. The Mothman soared away from the air base. Manfred's jaw clenched, willing at least one bullet to strike the beast.

It flapped its enormous wings and shot off into the night. A piercing wail washed over the base. A hellish sound that sent a chill up Manfred's spine. Again, it made him wonder if there was something to Erhardt's claim about these things being demons. What earthly creature could make such a blood-curdling noise?

He stared at the dead Mothman. Could a demon be felled by mere bullets?

"So they can make sounds," said Kretschmer.

"A sound I hope I never hear again," Erhardt spoke in a shaky voice.

"On that, I agree," Gude added.

Manfred rose from behind the machine gun just as Moritz trotted up next to him. He patted the great dane's head. "Good boy. You may have saved our lives."

He looked at the body of the Mothman, its torso ripped apart, blood pouring out the countless large holes. Manfred held his breath, thinking about the terror that had consumed him just minutes before. How could he be so scared? Like a small child?

No. Even as a small child he could not remember being scared like that. To the point it overwhelmed his body, his very soul. How could this be? How many battles had he fought in the air and on the ground since this war began? Even when faced with something as unusual as this Mothman, he should not be as frightened as a raw recruit in the middle of his first artillery barrage.

Kretschmer stepped up to the front of the sandbag emplacement and stared down at the dead Mothman.

"It seems you have gotten your wish, Doctor," said Manfred. "You have a Mothman to examine."

"*Ja.*" Kretschmer did not even look at Manfred, just gawked at the creature.

Manfred slid next to the doctor. "I want you to start your examination right away. Now that we know there is more than one Mothman, I want to know everything we can about these damn monsters."

NINETEEN

Manfred ordered the sentries doubled and everyone at the base to arm themselves. Unfortunately, they did not have enough rifles and pistols for all the men. Many of the mechanics, cooks, and other support personnel had to rely on wrenches, kitchen knives, or meat cleavers. His own mechanic, Ewers, talked about sawing off the leg of a chair and gluing nails to it. Manfred gave the man credit for his improvisation, but after what happened earlier tonight, he'd rather fight the Mothman from a distance with a rifle than up close with a piece of furniture.

We have to make do with whatever we have . . . for now.

Manfred continued writing at his desk, candles burning to either side of him, when Voss entered the tent.

"We have another problem, Manfred," said the younger pilot.

"Why does this not surprise me?" He snorted and leaned back in his seat. "What is it now?"

Voss settled onto the stool in front of the desk. "One of the mechanics, Private Löwith, has become hysterical. Like Private Schaffner, he's screaming and rambling about red eyes. Seeing the Mothman must have been too much for him."

Manfred just nodded. That made it three people the monster – *monsters* -- had turned into raving lunatics. Yes, it was terrifying. But as soldiers of the Imperial German Army . . .

Perhaps I should not be so quick to judge. He recalled his own horror in the machine gun nest. Scared to the point he could not move, could not think. Had it not been for Moritz . . .

He lowered his eyes, shoulders tensed. How embarrassing. He'd been fighting this war from the start, faced death hundreds of times. In all that time, his courage never faltered.

So why had it tonight?

"Manfred?"

Voss's voice pierced his thoughts. He stared up at his friend and said, "I take it Doctor Kretschmer is looking after him."

"One of his assistants is. The good doctor is busy cutting into that dead Mothman." Voss grunted out a laugh. "He's welcome to it. After what happened tonight, I'd prefer to never see those damn monsters again." He grimaced and turned his head to the side.

Manfred studied his friend's expression. He picked up a sense of shame. That's when he remembered how the other pilots in the machine gun nest reacted to the Mothman. All seemed frozen in fear like him. That surprised him. Especially with Voss. The younger man was one of the most daring and fearless pilots Manfred knew. How could he freeze like that in a fight? How could all of them, at the same time?

"I am sorry to say you will probably see them again," said Manfred.

A frown tugged at the corners of Voss's mouth. "How many of these things can there be?"

"I don't know. I do know that there are two less after today."

"Thanks to you."

Manfred nodded, tapping the paper on his desk with the tip of his pen. "At first light, I am sending a messenger to General von Armin requesting more troops, rifles, and machine guns to guard our base against any future Mothman attacks. Though given all the fighting along the Ypres Salient, there is a good chance he will deny my request."

"I wouldn't be so sure. Any request made by *Der Rote Kampfflieger* carries a lot of weight." Voss tacked on a grin.

"We shall see. Meanwhile . . ." Manfred set down his pen and folded his hands on the desk. "At dawn, I want you to take *Jasta* Ten and patrol a forty-kilometer radius around our base. It is possible tonight was a probing attack and the Mothmen are marshaling their forces nearby for a full-scale assault on us."

"We'll be in the air as soon as the sun comes up. Hopefully we will find them, unless they've gone back to Hell like Erhardt believes."

"Mm." Manfred's jaw stiffened, thinking about how he had to slap the pilot before his sanity completely slipped away.

He returned his attention to Voss. "Go back to your tent and get some rest. And keep your rifle close by in case the Mothmen return."

"Oh, I will keep it close by. As in I'll be holding it in my bed like it was my wife . . . if I had a wife."

Voss got up and headed for the tent flap when Manfred called to him.

"*Ja?*" Voss turned to him.

"Find *Leutnant* Erhardt and send him to me."

Voss acknowledged the order and left.

Manfred resumed writing his report. Ten minutes later, Erhardt entered the tent.

"You wished to see me, *Herr Rittmeister?*"

"Yes. Sit."

With a nervous grimace, Erhardt sat on the stool.

Letting out a long breath, Manfred leaned forward, locking eyes with the other pilot. Erhardt swallowed.

"What happened to you tonight?" Manfred demanded in a firm voice. "You were scared out of your wits when the Mothmen attacked."

Erhardt shifted in his seat and lowered his head.

"Look at me." Manfred snapped. "When one of my pilots loses his nerve like that, I want to know why."

"S-Sir, I . . ." Erhardt drew a shaky breath. "I just . . . the fear. It overwhelmed me."

"*Leutnant,* you have been in countless battles. You have a reputation of being a fearless pilot. Yet tonight you were almost frozen in place."

"It must have been the creatures. They were so --"

"That is no excuse. You are an officer in the Imperial German Army. You have been trained to fight and defeat any enemy with skill and courage, even if that enemy is something as unusual as these Mothmen. Anything less is unacceptable."

A flicker of shame went through Manfred. He felt like a hypocrite saying that, given how he could not move when the Mothman stood in front of him.

"I understand, *Herr Rittmeister.* But what happened . . ." Erhardt bit his lip. His shoulders rose, then stiffened. He opened his mouth with, to Manfred's perception, great effort.

"As I said, the fear overwhelmed me. I tried to fight it, I really did. But it was as though . . ."

Manfred tilted his head to the right. "As though what?"

Erhardt rested his hands on his lap and pressed his fingers together. "As though . . . something was making me be so afraid.

All the times I've been shot at in the air, I have never felt that way. But tonight . . . it was different. I could not stop being afraid. It swelled like air being pumped into a balloon."

Manfred cranked an eyebrow. He thought back to not only tonight, but when he and the Mothman dueled in the air. He could not remember ever experiencing such all-consuming fear, even in his very first battle as a cavalry officer. Given all the times he'd seen the monsters before tonight, why should he still be so afraid of them?

"They are demons." Erhardt interrupted his thoughts. "Perhaps they have some kind of power to make people very afraid."

"Enough of this talk of demons, *Leutnant.*" Manfred sliced a hand across the air. "I killed two of these things with ordinary machine guns. Could I have done that to an actual demon?"

Erhardt's mouth hung open for a moment. "I . . . I don't know. Maybe our weapons can kill them."

"Or maybe they do not come from Hell at all. Maybe General von Armin was right when he said the British or French found them in one of their colonies and released them in Flanders to attack and torment us. Whatever the case, there is a more logical explanation to their presence."

"What other explanation is there other than they are demons?"

"*Leutnant* Erhardt." Manfred's voice rose. "I am ordering you to cease this talk of demons."

"Sir, they are real," Erhardt blurted.

"*Leutnant,* I have given you an order."

"I have seen them before!"

Brow furrowed, Manfred held the other man in his gaze. "You have seen them before? How? When?"

Erhardt took a long breath before speaking. "My older sister, Claudia. She was kind, thoughtful, spirited. Everyone loved her. But when I was nine, something happened to her."

His shoulders slumped. "She started to argue with our parents. She yelled, she swore. Used words that stunned our parents. It was just . . . so sudden."

Erhardt stared up at Manfred, his jaw stiffened for a moment. "Then one day, our mother told her to set the table for lunch. Claudia began screaming at her, called her . . ." He winced. "Called her a fucking hag who should suck a goat's cock. Mother slapped her. Then . . . Then Claudia slapped Mother, so hard she knocked her to the floor."

Erhardt tensed, his gaze falling to the ground. "Then she yelled in this . . . evil voice. It sounded nothing like Claudia. She threatened to rip out our mother's throat. I was crying and she yelled at me in that voice, that she would gouge my eyes out and make me swallow them."

Manfred said nothing, just observed Erhardt. If the man was making up this story, then he would have a great career acting on stage after the war. But he doubted the pilot was putting on a show. Erhardt seemed genuinely upset telling this story.

"My parents locked her in her room as punishment," Erhardt continued. "She destroyed everything in it. Furniture, mirrors, clothes. We called a doctor, who told us we should send Claudia to an asylum. My father would not allow it. He did not want our family to be embarrassed in such a way. I remember living in fear every day that she would hurt me or my parents. Then my mother had our pastor come to the house to pray over Claudia. Ten minutes after he arrived, he ran out the door in a fright. We had two other pastors come over. One said she was simply an unruly child who needed more discipline."

Erhardt shook his head, face scrunched, his anger evident. "They must have seen what we all witnessed. The strange, almost roaring voice. Throwing furniture too heavy for a twelve-year-old girl to lift by herself. Finally, my mother invited a pastor from another town to come over. He determined that a demon had possessed my sister. How to deal with it became a problem. No senior members of the Protestant clergy would acknowledge the existence of demons. So the pastor decided to help on his own. He did research on possessions and exorcisms and came back to the house to get rid of the demon."

He shuddered. "It went on for three days. I . . . I saw candles and picture frames fall over on their own. One of our windows suddenly shattered. I . . ."

Erhardt clasped his hands together. Manfred watched the blood drain from the man's face. "I watched Claudia float off her bed and just hang in the air."

Manfred remained silent, trying to digest this story. He wanted to call it nonsense. But everything about Erhardt's reactions, the timbre of his voice, sounded so convincing. Had he actually seen this? Could his nine-year-old mind have misinterpreted what happened to his sister?

"The pastor did rid the demon from her, but . . . Claudia was never the same." Erhardt's face sank. "She became quiet, withdrawn. To this day she spends most of her time just staring out the window of her bedroom."

"I see." Manfred could not think of anything else to say. This was not the first time a pilot had come to him with a personal matter. But it usually involved a woman or some normal problem with their family. For this, he had no advice.

He took a slow breath, trying to give his brain time to come up with something. "I am sorry for what happened to your sister."

"Thank you, *Herr Rittmeister.*"

Unable to come up with any more words, Manfred said, "You may go, *Leutnant.* Get some rest . . . and perhaps pray."

"Yes, *Herr Rittmeister.*"

Erhardt rose slowly and exited the tent without a word.

Manfred stared in silence at the tent flap long after the other pilot left, the story echoing through his mind. Even if Erhardt's story was true, it did not mean these Mothmen were demons. Again, would machine guns kill a demon?

So if not demons, what then?

His thoughts switched back to *Leutnant* Erhardt. Whatever had happened to his sister when they were children had obviously scarred him. That brought about another concern.

Would that experience affect Erhardt in combat? And would it jeopardize everyone in the unit?

TWENTY

"Herr Freiherr. Herr Freiherr."

The voice, and the hand tapping his shoulder, pulled Manfred back to the world of consciousness. He forced his heavy eyes to crack open. A shadowy figure bent over his bed. He held his breath, fearing the Mothmen had returned. That's when he recognized the voice. Dr. Kretschmer.

"Doctor?" Manfred groaned. "What is it?" He blinked a couple of times, his darkened tent coming into focus.

"I've finished my examination of the Mothman you killed. I thought you'd want to hear what I learned as soon as possible."

I do, but perhaps after I have gotten more sleep. He had lain down in his cot not long after he'd finished his report to General von Armin. It was too dark to check his watch, but given the heaviness of his eyelids and the burning sensation of his eyes, he doubted that could have been too long ago.

But the doctor was right. He needed to learn all he could about this new enemy, and the sooner the better.

Manfred pushed himself off his cot and plodded to the stand in the corner of the tent where his wash basin sat. He poured out some water from the pitcher, scooped it in his hands, and splashed it on his face. The water was warm and barely did anything to bring him to full wakefulness, but it was better than nothing.

He followed Kretschmer out of the tent, the older man walking briskly. That surprised Manfred, since the staff surgeon had to have been up going on twenty-four hours. But he had spent a good portion of the night examining a creature unknown to the world until now. That probably made him ignore any fatigue he might have.

The pair entered the tent Kretschmer used for a makeshift morgue. Lit by candles, Manfred glanced at one table with a white sheet covering a human form. Probably Private Uhlig. On another table lay the dead Mothman.

Manfred grimaced at the creature, whose torso and skull were splayed open. Blood stained much of the table around the corpse.

"So, Doctor. What have you found out about this thing?"

Kretschmer stood beside the examination table. "As I suspected, the creature's physiology is not dissimilar to that of a man. Heart and lungs in the chest, stomach in the midsection. The muscles, though, are not as dense as I would have suspected for a creature this size. The bones are also hollow, much like a bird's. I imagine this aids the Mothman in flight."

The doctor walked to the edge of the table where a scale sat. Manfred eyed the roundish shape sitting in one of the bowls and grimaced. It had to be a brain.

"The shape and weight of the brain," Kretschmer pointed to the scale, "are different from that of a human one. For one, it is heavier. Whereas a normal human brain weighs around thirteen hundred grams, this one weighs close to fifteen hundred grams. It was also connected to the antennae on the top of its head."

"Do you have any idea what those antennae are used for?" asked Manfred.

Hands on his hips, Kretschmer stared at the Mothman's brain, jaw stiffening. The silence lingered for several seconds.

"Doctor?"

Kretschmer let out a long breath, but did not speak.

"You do have some theory, I take it?" asked Manfred.

"I do, though it will sound far-fetched."

"Look at what you have on your table." Manfred aimed a hand at the dead Mothman. "Everything about this is far-fetched. I am willing to entertain whatever theory you have concerning these monsters."

Kretschmer nodded. "Very well."

He walked around the table and approached Manfred. "I feel like I am seeing a pattern with these Mothmen. Specifically, how we have been reacting to them."

Manfred folded his arms across his chest as the staff surgeon continued. "Three men have been reduced to hysterics because of these creatures. Then I observed what happened when we fought this Mothman tonight."

Kretschmer bit his lip for a moment. "I mean no disrespect when I say this, *Herr Freiherr,* but you, Voss, Erhardt, Gude. In the machine gun nest, you were all frozen."

Manfred's cheek twitched. He did not want to be reminded of his failing that almost got them all killed.

"I was not immune, either. Looking at that creature, I found I could not move, much as my brain screamed at me to."

Kretschmer shook his head. "Combat is nothing new to me. I was assigned to field hospitals at Loos and the Somme. I treated wounded with shells exploding around me. Was I scared? Absolutely. I felt the urge to freeze, but I did not. I performed my duties. But when I encountered the Mothman, the fear I felt was . . . different. As though something magnified it."

Manfred's brow crinkled. He focused on the word *magnified.* That perfectly described every encounter he had with the Mothman. As though the usual fear he felt in battle, the fear he always tempered, was stoked like a fire. To the point he could not control it. That should not happen to someone who had seen combat as much as him.

But it did.

"Then there were its reactions," Kretschmer continued.

"What do you mean?"

"When the Mothman confronted us at the machine gun nest. The way it tilted its head back, nearly closed its eyes. I swear it looked as though it was in . . . ecstasy."

Manfred screwed up his face. "Ecstasy? You must be mistaken, Doctor. How can such a creature feel that kind of emotion?"

"That is my best interpretation of what I saw," said Kretschmer. "But there is also your account of your fight with the Mothman you shot down. You said it grabbed the sides of its head when it followed you close to the battlefield."

"Correct."

"You assumed it had been shot. But I believe its reaction was the result of something else."

"Which is?" Manfred cranked an eyebrow, wondering what the doctor was getting at.

Kretschmer raised a finger. "All these reports, all these observations, they lead me to believe that the Mothmen somehow amplify our fears. Not only amplify them, but I believe they actually feed on them."

Manfred said nothing, just stared at the staff surgeon. *Preposterous.* The word echoed in his head. But was it? Kretschmer's theory would explain the nearly uncontrollable terror he experienced whenever he confronted a Mothman. It also gave credence to how Erhardt described his reaction to the monster.

"But how? How can it do all this?"

Kretschmer stared at the ground in thought for a moment. "You've heard stories of mind readers."

"Yes, and I have always considered those who make such claims to be charlatans."

"As do I. But what if somehow these Mothmen have the ability to, well, not read our minds. But read our emotions. Influence them to their own ends."

"To feed on them." Manfred shook his head. "How can anything actually feed on fear, or any other emotion?"

"That I cannot answer." Kretschmer turned back to the dead Mothman. "I have only just begun to examine this creature, and it may take doctors and scientists smarter than me to understand how it can do such a thing."

"But you are convinced it somehow feeds on our fear."

Kretschmer nodded. "Yes. When we were all paralyzed at the machine gun nest, the Mothman could have attacked us at any time. But it did not. It just stood there. Why would it do that if it were not consuming our fear? Then there was the way the Mothman you fought reacted over the trenches. Tell me, what happens when you eat too much?"

"You get sick," answered Manfred.

"And that is what I think happened to that Mothman. It got too close to the trenches, where thousands of men were fighting and dying. Where thousands of men were scared they would be shot or blown up any moment."

Manfred's eyes widened as the realization hit him. "And all those men's fears were too much for it. It . . . became sick."

"Exactly." Kretschmer stabbed a finger in the air for emphasis. "It would seem the Mothmen can tolerate the fears of a small group of men. But a larger group, a regiment, for example. All that fear would overwhelm them."

He let out a relieved breath. "I guess we should consider ourselves fortunate. If these creatures could magnify the fears of thousands of soldiers at once, our entire front could collapse. Then the British and French could simply walk through Flanders."

Manfred grimaced. "I'm sorry, Doctor, but that is where you are wrong."

"How so?" Kretschmer tilted his head.

"The Mothmen do not need to affect an entire regiment to weaken our lines. Attacking small sections of our trenches could be enough to create a few gaps in the line. Gaps the Entente could exploit for a breakthrough."

Manfred gazed hard at the creature. "It is imperative we find these Mothmen and kill them. If we don't, we risk being pushed out of Belgium. Should that happen, nothing would stop the British and French from invading Germany itself."

TWENTY-ONE

Manfred studied the faces of his pilots at the morning briefing after telling them Kretschmer's theory about the Mothmen stoking and feasting on people's fears. Some looked skeptical, some accepted it, some appeared to mull it over. A few simply looked tired. Understandable after last night's attack. They probably had trouble sleeping.

Even Manfred had to fight to keep his eyes open. Not even breakfast and two cups of coffee could completely fend off his tiredness. But he could not nap here in the mess tent in front of his pups. He'd have to will his way through this briefing, and the rest of the day.

Hans Adam, who sat with his face scrunched in disbelief, raised his hand. "Are we certain about this? A monster that . . . eats fear?"

Manfred turned to Dr. Kretschmer, who stood nearby. The staff surgeon gave a small shrug. "I do not have any proof of this, but I feel it is likely based on *Rittmeister* von Richthofen's account and what I have seen of the Mothmen myself."

He stepped closer to the pilots. "Think back to last night. Did you have an intense fear? A fear none of you had ever experienced, even in battle? That was probably due to the Mothmen's ability to enhance any fear you already had."

The men of *Jagdgeschwader 1* remained silent. A few turned their gazes to the side or the ground. Manfred doubted any of them would admit to being scared. Not these men who faced bullets in the sky on an almost daily basis.

"I know I felt . . . strange." *Leutnant* Erhardt broke the silence.

All eyes turned to him. The veins in his neck stuck out. "I could not move when that demon stood in front of me. All the battles I have been in. All the times I have been shot at. I did not flinch. I fought, every time. But last night . . . to be frozen in the face of an enemy?"

Erhardt shook his head. "No. That should not have happened. The demon must have made me afraid. That is what I would expect from a minion of Lucifer."

Several men cranked eyebrows or tilted their heads as they stared at Erhardt. Manfred did not. He kept his face stiff. Erhardt still insisted the Mothmen were demons from hell. Manfred was not so sure. But his wingman had admitted to the overwhelming sense of fear when confronted by the monster, similar to what he experienced.

Perhaps . . .

"I had the same feelings as *Leutnant* Erhardt."

Manfred's head jerked to the left, stunned by the voice that spoke those words. Of all the men under his command, he never imagined Werner Voss would share his fears with his fellow pilots.

All eyes turned to the leader of *Jasta* 10, many wide with surprise.

"I have never cowered before the enemy," Voss continued. "Never stood like a statue to give another the chance to kill me. But that is what I did last night." He lowered his gaze, scowling. "On my honor, any other time I would have brought up my pistol and shot that bastard in its ugly face. So it must have . . . affected me to turn me into a terrified child."

Three more pilots spoke up about the all-consuming fear they experienced during last night's attack.

"For a second," said Gude, "I thought I would start screaming like poor Förster. Thankfully, I did not."

"Then not only does it appear Doctor Kretschmer's theory is correct," said Manfred. "It seems we may also be able to resist the Mothmen's ability."

Hands clasped behind his back, he continued, "In spite of the fear these monsters created in us, we did not go into hysterics."

"If I may, *Herr Freiherr.*" Kretschmer held up a finger.

Manfred nodded and the doctor spoke. "One possible explanation of Förster's reaction is the Mothman he encountered was unlike anything he had ever seen, anything he could have imagined. A creature, as *Leutnant* Erhardt says, straight from the pits of Hell. That, and its ability to manipulate our fear, may have proved too much for Förster."

A small grin formed on the doctor's face. "But now we know the Mothmen exist. They should no longer be mysterious and terrifying to us. Whenever your fear begins to consume you, you must tell yourself it is coming from the Mothmen, not you. Though it will likely require great willpower by all of you."

"Then we shall exercise great willpower, Doctor," Manfred added in a firm, confident tone. "We will not lose this battle because we are weak of mind."

Manfred ran down the flying assignments for each of the *Jastas,* along with the two CIII reconnaissance planes. The men then headed out of the mess tent to change into their flight gear.

Walking toward his tent, Manfred scanned the sky. A few dark clouds hung in the distance. He frowned. It seemed more rain was in their future.

"So you really think what Doctor Kretschmer says is true?"

Manfred turned to find Voss behind him. "Do you?" he asked the younger pilot.

Voss gave a slight shrug. "I guess I have to after what we experienced last night. But it still sounds so unbelievable."

Letting out a short sigh, Manfred said, "I am still struggling to accept what the doctor told us. How is it possible for anything to . . . reach into our minds, and turn it into a weapon against us?"

Voss looked away, jaw stiff, as though thinking. A few seconds passed before he spoke. "I now wonder if *Leutnant* Erhardt was right all along."

"About?"

Voss stared him right in the eyes, his expression more serious than Manfred had ever seen. "That these Mothmen are actually demons."

***

That statement from Voss clung to Manfred's mind throughout his patrol. He had been dismissive of Erhardt's claims for so long, even after he confided in him about his sister's possession. Now . . . well, he couldn't say he believed it as fervently as Erhardt, but he was more willing to accept the possibility than he had been just a day ago.

Demon or not, at least bullets can kill it.

Manfred and his squadron continued their flight over Flanders, flying lower than he would have liked. Unavoidable if they wanted to spot the Mothmen in their nest, or wherever they resided when not flying.

When not eyeing the ground, Manfred gazed north. Apprehension knotted his shoulders as the dark gray storm clouds loomed before him. He thought back to the flight back to base after retrieving the camera from the downed CIII. Rain pounding him and his Albatros, barely able to see more than a few feet beyond his propellers. He had been fortunate to make it back to base. He'd rather not go through that again.

Other grayish clouds floated on the horizon to the south. Smoke from German and British artillery battering each other up and down the Ypres Salient. Manfred frowned, thinking of the soldiers who will have to fight in trenches and fields turned into a muddy soup by the coming rains. He'd gone through that a few times as a cavalry officer. Soaked, filthy, chilled to the bone, bullets whizzing over his head, shells exploding nearby.

Using the word "miserable" to describe such conditions was inadequate. Manfred doubted any words existed to truly capture the wretchedness and horror of such a battlefield.

They found no sign of the Mothmen before having to return to base. Manfred was planning a second patrol when rain hammered the airfield. The storm lasted into the night. Light showers followed the next day, but the low-hanging gray clouds over Flanders grounded *Jagdgeschwader 1*.

Manfred scowled at the darkened sky. Did the storm ground the Mothmen as well, or could they navigate through it without trouble? He did not like the thought of his unit sitting idle while those monsters could be gathering for another attack. The aggressive part of him wanted to order his planes in the air, risks be damned. The rational part of him dismissed that idea. He'd lost a few planes and pilots flying through the last storm. He would not let the elements take more of his pups.

Everyone made the most of their forced time on the ground. The mechanics performed much needed maintenance on the planes. The photographic interpreters examined each picture from the CIIIs multiple times, looking for any sign of Mothmen. They found none.

Manfred, meanwhile, caught up on the bane of every commander's existence . . . paperwork. Requisition forms for everything from bullets to spare parts to food. Updates to General von Armin on their search for the monsters – not that there was much to report, thanks to the thunderstorms. A transfer request from one of the mechanics to join the Air Service. Pilot evaluations from his *Jasta* commanders.

Except Voss.

Manfred sighed in frustration. It did not come as that much of a surprise. Voss's disdain for paperwork was no secret. But being a *Jasta* commander, he needed to take the role more seriously. To sit at his desk and write his evaluations and combat reports instead of chatting with the mechanics or riding around on his motorcycle.

It is time to learn some responsibility, my friend.

But he would do that tomorrow. Today was reserved for finishing paperwork. Tonight would be spent inspecting the base defenses, just in case the Mothmen returned.

***

Cold wind blasted over Manfred's head and shoulders as his Albatros D.V. soared above Flanders. The C.III droned on to his right, though he paid scant attention to it. Most of his attention was on the skies around him. His job, and that of Erhardt's, was to protect the reconnaissance plane while it searched for the Mothmen's nest. Not only from the winged creatures, but from any British or Commonwealth fighters.

There still is a war on.

In Manfred's mind, there were two wars. The one he'd been fighting for three years against his human enemies, and the one against the Mothmen. After the attack on his base a few nights ago, he could not imagine them serving the British, the French, or any other nation. The Mothmen most likely served the Mothmen.

Manfred swung his head left to right, then checked over his shoulder. No threats. He also looked to the ground, in case the creatures might be flying up to engage them. He spotted the ruins of a small village called Kippe, but no flying monsters.

The observer in the C.III twisted in his rear seat, then leaned over the side of the plane, bulky camera in hand. Manfred hoped the pictures showed some evidence of the Mothmen's nest.

The reconnaissance plane flew over a few more villages and an abandoned trenchline before the pilot signaled he was getting low on fuel. The three-plane formation swung around and headed east. It would now be up to the photographic analysts back at base to see if the camera picked up anything of note. Manfred tried to be optimistic, but had his doubts. On a map, Belgium was tiny compared to Germany and France. In reality, there were thousands upon thousands of square kilometers for the Mothmen to hide.

The battle for the Ypres Salient could be over by the time we find them . . . if we even do.

Manfred grimaced. Would the battle end in favor of Germany or Britain? And would the Mothmen have any influence on the outcome?

He kept watching the skies. No other planes – or Mothmen – were in sight. Manfred turned his gaze to the ground. This time he did spot something.

A line of trucks rolled along a dirt road, approaching a river. Likely carrying troops or supplies, or both, to the front. A few tents sat on both sides of the bridge, enough to accommodate a platoon of soldiers. Parked along the south side of the bridge was a truck with a large gun mounted on its bed. An M1914. A good location for the mobile anti-aircraft gun. The bridge would be a prime target for British bombers to try and disrupt their supply lines.

The planes made it back to base without incident. Sharp pain pierced Manfred's skull soon after he landed. His stomach burned and swirled. He gritted his teeth, breathing deep through his nose. He gripped his knees and shut his eyes. *Go away. Go away.*

"Herr Rittmeister?"

Manfred opened his eyes. Ewers stood beside the cockpit. The young man bit his lip in a worried expression.

"Just taking a quick rest." He spoke the words quickly, afraid if he kept his mouth open longer he would not be able to hold back his nausea.

Grimacing, Manfred pulled himself out of the Albatros and eased himself to the ground. His legs quivered, but he forced himself to stay upright. He glanced at Ewers. The worry remained on the mechanic's face.

"Check it over." Manfred jerked his head back toward his Albatros. The gesture sent an invisible fist smashing into his brain. He fought off a dizzy spell and said, "Refuel it. I will likely be flying again soon."

"Yes, *Herr Rittmeister.*"

Manfred ignored Ewers' concerned gaze and trudged back to his tent. As soon as he entered, he dropped into his cot. There he lay for an hour until the headache and nausea subsided. He got up, threw water on his face from the wash basin, and sat at his desk. In his mind, he made a mental list of all the non-flying tasks he needed to get done today. One of them was a meeting with Voss. But his friend was leading *Jasta* 10 on a patrol over Passchendaele, so that would have to wait.

He was writing his report on this morning's mission when Alois Heldmann appeared at the tent's entrance.

"Herr Rittmeister. Sorry to interrupt, but a Major Brühl from Fourth Army staff is here to see you."

Manfred sighed. *This is sure to be a waste of time.* He nodded and stood.

Heldmann stepped aside, and a lean, bespectacled man wearing a gray tunic, polished black boots, and a peaked hat walked in. He stopped by the stool, the skin around his nose crinkling.

Probably used to sitting in actual chairs, thought Manfred.

Brühl let out a brief snort and turned to him. *"Rittmeister* von Richthofen. General von Armin has sent me here to get a progress report on your efforts to find these . . . Mothmen."

"Of course, sir."

Brühl aimed a hand at Manfred's chair, indicating for him to sit. He did so. The major then stared at the stool again, frowning. He removed a handkerchief from his breast pocket and wiped it down.

Manfred suppressed a sigh. Brühl had to be a career staff officer. Anyone that concerned about a little dirt on a stool probably never spent one day in a trench. His dislike for the man was immediate.

"So . . ." Brühl took out a small journal and a pencil from another pocket on his tunic. "Have you been able to locate these . . . creatures?"

"Not yet. The storms of the past couple of days have kept us grounded. We are concentrating our searches on some of the devastated villages in this part of Flanders. They are, for the most part, abandoned. We do not even have garrison troops there. They would make perfect nesting sites for the Mothmen."

Brühl responded with a soft grunt. *"Rittmeister,* it is imperative you and your pilots find these creatures."

Manfred's eyebrows knitted together. *Do you think I am not aware of that?* His lips cracked open, releasing a breath, expelling at least some of his anger at the major.

Brühl continued, "The British and French have already captured Pilckem Ridge. Our intelligence shows British forces gathering near the Gheluvelt Plateau. We expect them to launch a major attack any day. Entente fighters are gaining the upper hand in the air, while our best pilots are distracted hunting for monsters."

This time, Manfred did not bother hiding his ire for this man. Eyes narrowed, he spoke in a deliberate tone. "I assure you, Major, we are not *distracted.* We are following the orders given to us by General von Armin himself."

"Then you need to follow those orders with haste. Your squadrons are desperately needed over the Ypres Salient. If the Entente breaks through our lines, they could threaten our U-boat bases along the Belgian coast."

"Even with our planes, there is much ground for us to cover. It is also likely the Mothmen are nesting in buildings or in forests, hidden from our eyes. If you wish us to find them quickly, we need soldiers on the ground searching for them. At least a division, maybe two."

Brühl's eyes widened behind his glasses. "A division? *Rittmeister,* the Entente is attempting a breakthrough at the Ypres Salient. We cannot afford to spare a division to poke through trees and bombed out buildings."

"Then perhaps a regiment," said Manfred.

An exasperated breath flowed from the major's mouth. "*Rittmeister,* we cannot give you a regiment, or a company, or even a platoon. Every soldier is needed to hold the salient."

Manfred leaned back in his chair. "Does that mean you cannot even spare a handful of men to augment the security of this base? Which was attacked by the Mothmen just a few nights ago."

Brühl gave a short shake of his head. "Again. All our soldiers are needed to defend the Ypres Salient. If you need more sentries, I suggest arming your cooks and mechanics and supply personnel and have them walk the perimeter."

Manfred aimed his eyes at Brühl's glasses. So that was it. General von Armin had impressed upon him the importance of finding the Mothmen, but would not commit any more forces to that effort.

It also made him wonder. If 4th Army could not give him even a couple dozen soldiers to aid in their mission, just how desperate was the situation in Ypres?

Major Brühl then asked for all of Manfred's reports pertaining to the Mothmen. These included photographs of the creature they killed at the base and the notes of Dr. Kretschmer's autopsy.

Holding the staff surgeon's papers in his hand, Brühl tilted his head. "Your doctor claims the Mothmen . . ." His face scrunched in disbelief, "possess an ability to mentally enhance a person's fear . . . and feed on it?"

He looked up from the report. "How is that possible?"

"We do not know, *Herr Major,*" replied Manfred. "At this point, it is a theory of Dr. Kretschmer's."

"And this Doctor Kretschmer, would you consider him a . . . crackpot?"

Manfred scowled. "The doctor has a . . . unique way of looking at things. But he is highly intelligent and has a very analytical mind. In his time serving *Jagdgeschwader* One, I have come to trust his judgment."

Brühl grunted, unconvinced.

The meeting went on, and on, to the point Manfred had to send a mechanic to the mess tent to bring them lunch. At the sight of the potato stew and piece of bread, Brühl sneered. Probably used to having his food prepared by whatever chef General von Armin had back at his headquarters.

The major wanted details of the attack on the base, what abilities the Mothmen had. Nothing to do with manipulating someone's fears. But strength, sharpness of their claws. Could they communicate with each other? How fast could they fly?

When they talked about Manfred's search patterns, Brühl suggested they expand their patrol radius into Entente territory. "That is where they probably launch their attacks from, since they are serving the British and French."

"You believe they are?"

"That is what General von Armin believes."

Manfred's lips tightened. He had a hard time believing that. But he kept that to himself. Even his reputation wouldn't save him from the consequences of telling a general he was wrong.

It was late afternoon by the time Brühl finished his meeting and got back in his staff car for the return trip to 4th Army Headquarters, taking all reports and photographs on the Mothmen with him. Manfred scowled as he watched the car drive off. So many hours wasted. Hours he devoted to more productive endeavors.

Like finding the Mothmen . . . quickly, like you want, he directed that thought at Major Brühl.

He headed back to his tent, coming across Gude along the way.

"Go find *Leutnant* Voss and tell him to report to my tent."

Ten minutes later, the commander of *Jasta* 10 slipped through the tent flap. Manfred cranked an eyebrow when he noted his friend's appearance. He was dressed in overalls, with grease stains on his left sleeve and shoulder. No doubt spending time with the mechanics, as he usually did when not flying. Any visitor to the base who took one look at Voss would never believe him to be an accomplished fighter pilot, let alone a squadron leader.

"You wanted to see me, *Herr Rittmeister?*"

"Yes. There are some things I have been meaning to talk to you about. Such as ---"

"Herr Rittmeister!" Someone shouted outside the tent. *"Herr Rittmeister!"*

A lanky man poked his head into the tent. A sergeant with the photographic interpretation unit.

"What?" Manfred snapped.

The sergeant swallowed. "Forgive the interruption, sir. But we found something in one of the photographs from today. You need to see it."

TWENTY-TWO

Manfred peered through the magnifying glass. There was no mistaking the form he saw in the photograph. Those wings could only belong to a Mothman.

Voss, who stood next to him, barked out a triumphant laugh. "Ha! We finally found the ugly bastards." He clutched Manfred's shoulder.

"It appears so." Manfred continued staring at the black and white image of the Mothman as it flew toward the bell tower of the church in Kippe, one of the few buildings left standing in the tiny village. It made the perfect roost for any winged creature, be it a pigeon or a beast like this.

Manfred set down the magnifying glass and straightened. He gazed around at the men in the photographic interpretation tent. "Good work, all of you."

The soldiers thanked him, a few beaming, no doubt ecstatic at being complimented by *Der Rote Kampfflieger*.

He glanced down at the photograph, wishing Major Brühl was still here. Manfred would have loved to hold the picture up to the staff officer's face and remove any doubt in the man's mind that he and his fliers were not performing their duty.

"Now that we know where they are, let's go kill them," said Voss.

Manfred checked his watch. His brow wrinkled as he ran some calculations in his head. "By the time we plan the mission, arm and fuel our planes, it will be close to dark. No, we will set out first thing in the morning."

He took the photographs with him, along with maps of the area, and returned to his tent. There he sat at his desk, studying the church, examining the surrounding terrain, deciding how many of his men to commit, and the best avenues of attack.

By the time the sun went down, Manfred had come up with his plan. He leaned back in his seat, staring at the papers, maps, and pictures cluttering his desk, and nodded. By this time tomorrow, the Mothmen would no longer be a threat to the German Army.

***

Manfred's skin prickled as his Albatros D.V. bore through the morning sky over Flanders. He took deep breaths, trying to settle his anxiousness. That proved futile. All he could think of was reaching Kippe and finishing off these monsters.

He sneered as the cold wind blasted over him. The soldiers and pilots who'd been killed or driven mad by these beasts would soon be avenged. Though he prided himself on never making fights personal, Manfred could not help but make an exception this time. The Mothmen had made him feel crippling terror, planted doubt in him about not just his manhood, but his very sanity.

They came close to taking away everything that mattered to him. Now he would take away their lives.

A rectangular object appeared on the horizon. Manfred's shoulders rose in anticipation. The bell tower of the church. The nest of the Mothmen.

He swung his head left to right. Two four-plane formations flew on either side of him. The one on the left led by Eduard Dostler, the commander of *Jasta* 6, the one on the right led by Voss. Three more planes trailed behind Manfred's Albatros.

The church grew larger. Manfred checked all around him. He anticipated the Mothmen to have sentries around the village. So far, the only thing in the air were his fighters.

The squadron soon reached Kippe, or what remained of it. Charred rubble lined the streets. Only a few structures besides the church remained standing. Even the church itself suffered from past battles. A portion of the rear had collapsed, and a large hole scarred the roof.

The planes circled the village, looking for any Mothmen. None were visible.

Sleeping in the bell tower? Manfred stared down at the pointed structure. Maybe the droning of their engines had awakened them.

Then we shall make it a rude awakening.

He turned to Dostler's Albatros D.V. and dipped his nose twice. The former Bavarian Army pioneer waved to him, waggled his wings, then dove at the church. The other three fighters in his charge followed.

Manfred, meanwhile, swung his plane east. His formation followed, as did Voss's. He checked over his shoulder in time to see Dostler rake the bell tower with his machine guns. Bits of masonry and wood jumped from the structure.

Manfred grimaced. Firing on a church felt sacrilegious. But they had no choice if they wanted to kill the Mothmen. And if Erhardt was correct and they were demons, then they were defiling the church. He was certain God would understand their actions as they were ridding His house of unholy intruders.

Dostler pulled up. Seconds later another plane strafed the bell tower, then a third. By the time the fourth fighter came in for its attack, Manfred was descending toward a field on the outskirts of Kippe. He and the other planes set down on the expanse of grass. It would not be enough to blast the Mothmen from the air. They needed to go into the church to confirm the monsters were all dead. Since 4th Army would not send him even a handful of soldiers, his pilots would have to do it themselves.

The Albatros rolled to a stop, its engine winding down. Manfred leaned forward, shutting his eyes tight to fight the pounding in his skull.

Be strong. You have a fight ahead of you.

He drew deep breaths, gritting his teeth as he straightened in the cockpit. Exhaling loudly, he grabbed the Gewehr 98 propped against the right side of his seat. Next, he got his old cavalry sword resting against the left side of his seat. Manfred had only ever used it for ceremonial occasions. But with the possibility of close quarters fighting in the church, the sword might actually serve him better than the rifle.

He leaped from the cockpit, legs trembling when he hit the ground. Manfred leaned against the Albatros to keep his balance. After a couple of breaths, he strapped his sword belt around his waist. Manfred then checked his other weapons. Service pistol, hand grenades, combat knife. All there. The other pilots were similarly armed. He had no idea how many Mothmen they might encounter. Best to bring as many weapons to this fight as possible.

The others gathered around his plane and split into two squads. Manfred with Erhardt, Gude, and Heldmann, and Voss with Adam and the remaining two pilots, Lieber and Plass. They set off for

Kippe, two kilometers away. Dostler's formation circled the village to cover them from any Mothmen . . . or Entente fighters.

Manfred's head throbbed as they walked across the field. Thankfully, his legs no longer felt like rubber. He fought through the pain in his head, reminding himself to keep an eye out for any Mothmen. The monsters were nowhere in sight.

They reached Kippe without problem. They sprinted from one rubble pile to another as they closed in on the church. Manfred wondered if it even mattered. If any Mothmen survived Dostler's strafing, they could see them easily from the bell tower.

Still they ducked behind scorched mounds of brick and wood as they neared their target. Manfred's eyes constantly darted from the ground to the bell tower, half-expecting one of the creatures to dive on them.

None did.

Manfred led his squad behind the remains of a wall. Voss and the others took cover behind rubble on the opposite side of the street. Across the way stood the church. Manfred took out his binoculars and looked over the bell tower. No movement.

"Gude. Heldmann. Stay here and cover the church, just in case any surviving Mothmen try and escape."

Both men acknowledged the order.

Manfred ordered the rest to fix bayonets. As with his sword, the blades at the end of their rifles might be more effective in an up-close fight.

The six pilots charged toward the church. Manfred glanced up at the bell tower. No Mothmen appeared.

They pounded up the stone stairs, which were chipped and pockmarked. Shrapnel hits most likely. They reached the heavy wood doors, also scarred from previous battles. Manfred nodded to Voss, who shoved open the right side door.

The men rushed inside, rifles up. Manfred looked above them. Nothing.

They moved through the church, but found no sign of flying monsters. Manfred then turned to the steps leading up to the bell tower. His heart pounded. Fighting in such a confined space would be difficult. But they had no choice. They had to make certain the Mothmen were dead . . . or kill any who survived Dostler's attack.

"Lieber. Plass. Stay here and cover our rear. The rest of you with me."

He noticed Erhardt swallow. Voss and Adam were stony-faced, but Manfred sensed the apprehension radiating from their bodies. He sympathized, looking back at the staircase. After fighting in the air for so long, this would be like brawling in a closet.

Exhaling, Manfred led the men up the stairs. He took it slowly, eyes focused straight ahead. If any Mothmen remained, they must have seen them enter the church. Perhaps they were waiting to ambush them at the top, to give them as little room as possible to fight.

Sweat broke out on his forehead. Manfred glanced at the bayonet, the muscles in his arms coiling, ready to thrust if one of the beasts appeared.

He made it halfway up the stairs. The door to the bell tower lay ahead. His breathing increased with each step he took. Any second he expected a Mothman to burst into view.

Ten steps to go. The world condensed to the door and the top of the staircase. It remained closed. Thin rays of sunlight flowed through the bullet holes from the strafing attack. Part of him wished the monster would just throw open the door and charge them, end this heavy dread that blanketed him.

Manfred reached the top of the staircase. He looked behind him and nodded. Erhardt, Voss, and Adam all gave a hesitant nod.

Turning back to the door, Manfred breathed deep, held it, and kicked it in.

He rushed inside, bayonet extended. The others ran in. Manfred swung his head left to right. Nothing. He checked above him. Nothing. He lowered his head.

That's when he saw it.

"Dammit," he hissed.

Voss stepped up next to him, gazing at the floor. "Well, at least we've kept the local mice population safe."

Manfred did not laugh. He just scowled at the torn, bloodied form of the owl laying at his feet.

TWENTY-THREE

Manfred sat at his desk, pen in hand, squeezing it over and over again. A way to try and purge the anger that had boiled within him since he'd stomped out of the bell tower in Kippe.

It did not work. His anger remained. In fact, it grew. So did his embarrassment. How could he have been fooled so badly? How could he have mistaken a damn owl for a Mothman?

The wings do look similar. From that altitude, it could be hard to judge its actual size.

That was a reasonable explanation. But reason be damned. He had made a blunder of epic proportions. One that a cadet at any military academy would not have made.

What made it worse was he'd have to write a report about it and submit it to 4th Army headquarters. What would General von Armin think when he read it? Would he doubt his ability to lead the best unit in the Air Service?

Manfred squeezed his pen harder, nearly piercing the skin of his palm with his fingernails.

"Herr Freiherr."

He looked to the tent flap. Dr. Kretschmer stood there, hands behind his back. "A minute?"

Manfred nodded, still gripping his pen.

Kretschmer entered. Moritz, lying next to Manfred's desk, looked up and wagged his tail as the doctor revealed his hands, and the contents they held. Two metal cups and a bottle of Schnapps.

"I heard about your mission in Kippe. I imagine that had to be disappointing."

Manfred scowled. "'Disappointing' does not come close to describing it."

Kretschmer grunted in acknowledgment. "I suppose you are right . . . and I suppose you need something from my medicinal cabinet."

Eyeing the bottle, Manfred's lips twitched in the briefest of smiles. He held out his free hand toward the desk. Kretschmer set down the cups and poured. Manfred took one and waited for the doctor to sit before taking a gulp. The strong, fruit-flavored liquid flowed down his throat. His tension melted away . . . some of it anyway.

He set down the cup and shook his head. "An owl. I actually thought an owl was one of those damned monsters."

"Well, when you're taking a picture from a thousand or so meters in the air . . ."

Manfred snorted. "That is no excuse for such an error in judgment."

Kretschmer stared at his cup in thought. "Or it could be your brain playing tricks on you. These Mothmen have been your main focus over the past couple of weeks. It might be possible you see some suspicious shape in a photograph and convince yourself it must be a Mothman. You and the men in the photographic interpretation section. You want to find it so bad your mind will turn anything into your quarry."

Manfred let out a small snort and sipped his Schnapps. He lowered the cup, but stopped halfway to the desk. His brow furrowed, thinking.

A few moments passed before he looked up at Kretschmer. "You said the Mothmen are able to manipulate our fear. Could they affect our minds in other ways? Make us think they are somewhere that they really aren't?"

The doctor leaned back, face scrunched as if considering Manfred's theory. "Perhaps. But that is speculation."

"Almost everything regarding the Mothmen is speculation."

"True." Kretschmer took a swig from his cup.

Manfred put down his drink and rested his clasped hands on the desk. "Can you do another examination on the dead Mothman? Maybe find out more about how it can affect our minds?"

Kretschmer frowned. "Truthfully, *Herr Freiherr,* I think I have done all I can when it comes to examining that corpse. I am equipped to treat wounded soldiers, not study a creature the world has never seen before. A university laboratory back in Germany would be better suited for that task."

"I can put in a request to General von Armin to transport it to a laboratory for a proper examination. Though I would not get your hopes up, Doctor. If he will not send a few soldiers to defend this base, getting a truck may be just as difficult."

"If the general feels finding and destroying the Mothmen is so important, I'm sure he can spare one truck."

Manfred gave a slight shrug. "You would think so. But it seems that, except for us, every soldier and vehicle in Fourth Army is committed to the battle at the Ypres Salient."

"Perhaps it will take another attack for the general to change his mind," said Kretschmer.

"Let us hope it does not come to that."

"Agreed."

The two men finished their Schnapps. Kretschmer collected Manfred's cup and stood. "I suppose I will look over the Mothman corpse again. If we wait for the Army to do something about it, the damn thing will have decomposed."

Manfred grinned at the joke.

Kretschmer stepped toward the exit when Manfred said, "Doctor."

"Yes, *Herr Freiherr?*"

"I want to commend you for your work during this . . . unique situation. Not just when it comes to the Mothmen, but . . . everything else."

"I am always at your service, *Herr Freiherr.*" A smiling Kretschmer gave a slight bow.

Manfred nodded back. With his brother Lothar still convalescing, the doctor was probably the only man on this base he could be somewhat open with.

Kretschmer exited the tent. Manfred sat at his desk, staring at the blank paper before him. In better spirits than he was a few minutes before, he decided to put off the report on the Kippe mission until later.

It seems some of Voss's bad habits are rubbing off on me.

After dinner, Manfred returned to his tent to try and write his report. He allowed himself to be distracted by Gude's nightly violin concert. Tonight was Brahms's "Hungarian Dance." The fast, spirited playing made it easy for him to again put aside writing about his failure and lose himself in the music.

You can only put this off for so long. He grunted a small laugh, wondering if he'd start to become lax in his paperwork, to the point his desk would resemble Voss's with stacks of unfinished –

A rifle cracked.

Manfred shot out of his seat as another shot split the air. Then another.

Someone yelled, "Mothmen! Mothmen!"

TWENTY-FOUR

Manfred raced around his desk and snatched the rifle leaning against his cot. He quickly slid on the bayonet and dashed outside. Mortiz barked, the Great Dane's head aimed at the sky. Manfred looked up. He made out one winged silhouette soaring over the base. Dammit, they needed flares to . . .

A streak of white rose from the ground, followed by another, and another. Seconds later three miniature white suns blossomed above him.

Mein Gott. Manfred gaped.

At least a dozen Mothmen circled the base.

Rifle fire rippled from one end of the installation to the other. A machine gun rattled. One of the monsters jerked, but remained in the air.

Manfred brought the Gewehr 98 to his shoulder and fired at the nearest Mothman. He pulled back the bolt and fired again. The creature kept flying.

Manfred hurried across the field, Moritz running beside him. He glanced from sky to ground, searching for pilots to gather into rifle squads to better defend the base.

The first one he came across was Voss. The young ace was on one knee firing at the Mothmen.

"Werner!" Manfred shouted.

Voss turned to him.

"Gather more men and protect our planes."

"Ja, Herr Rittmeister."

Voss leapt to both feet and took off running.

In the distance, a Mothman dove. So did another. Rifles and machine guns hammered away. One of the creatures went into spasms and fell from the sky.

Manfred aimed at another Mothman and fired once, twice, three times. He missed with every shot.

"Dammit." He dug into his pocket for another stripper clip.

"Herr Rittmeister."

He spun around to see *Unteroffizier* Ewers pounding toward him. He skidded to a halt a couple of meters away, breathing hard. Manfred glanced at the mechanic's right hand. He held the leg of a wooden chair with nails poking out of it.

"That is all you have?" He nodded at the improvised weapon.

"They did not have enough guns to give me one. I wanted something to use in case those monsters returned."

"Then it will have to do. Follow me." Manfred shoved the clip into his Gewehr and broke into a jog. "We need to gather more men and --"

A Mothman streaked over the ground, targeting one of the machine gun nests twenty yards away. Manfred raised his rifle and fired. Missed. His gaze shifted to the two men at the weapon, firing it into the air. Their backs turned to the threat.

"Behind you!" he shouted.

Neither soldier reacted. Distance and the chatter of the machine gun likely prevented them from hearing him.

Manfred pulled back the rifle's bolt and aimed.

The Mothman slammed into the soldiers. Both went sprawling to the ground. The MG 08 spiraled off its mount. The creature landed next to the fallen weapon. Manfred took aim.

A woosh of wind surrounded him. Someone screamed.

He spun around. "Ewers!"

The mechanic rose into the air, a Mothman clutching him under the arms. He cried out, kicking his legs. "Help! Help!"

The creature flapped its wings, climbing higher.

Manfred raised his rifle. Shadows closed around the base. The flares were fading. Darkness enveloped both the Mothman and Ewers.

"Dammit." He had no clear shot.

Ewers kept screaming. Manfred's throat clenched. The man who took care of his plane, who always showed concern for his well-being . . . and he could do nothing to help him.

A burst of white light burned away the night. Someone launched a new flare.

They did it just in time for Manfred to watch Ewers plummet toward earth.

"No!"

Mouth agape, all he could do was watch the mechanic wail in terror until he struck the ground. Manfred rushed over to the unmoving form, a hitch in his throat. Falling from that height, there was no hope for him.

He stopped a meter from the body and looked up. The Mothman circled above, head leaned back as if in ecstasy.

"Bastard!" He fired. The creature's leg jerked. It screeched and soared away. Manfred shot at it again, but missed.

Another scream caught his attention. Another man fell from the sky, arms flailing.

Manfred spotted a Mothman carrying someone else aloft. The man's legs kicked wildly until the creature let him go. A prolonged, terrified scream followed until he slammed into the ground.

Feeding on our fear. That had to be why the Mothmen did not kill them outright. They wanted to make those men feel helpless. To make them know death was imminent and they could do nothing about it. A veritable feast of fear.

It is time we make them afraid of us. He spotted a Mothman diving toward the base and tracked it with his rifle.

Wait . . . Wait . . . N—

Something moved in his peripheral vision. Manfred turned . . . then dropped to the ground.

A Mothman soared over him. He pushed himself to one knee just as it landed. The beast spun around and aimed its blazing red eyes at him.

Dread welled up in Manfred. His arms refused to bring up his rifle. All he could do was stare at those eyes. Eyes with no pupils. Just two red circles, so intense they could be made from the fires of Hell. Like a demon. Like what Erhardt believed.

It's not you. It's this thing. I am not afraid.

The Mothman stepped toward him. In its hands was a tubular machine gun, held the way an Englishman might hold a cricket bat. This must have been the same creature that attacked the nearby machine gun nest. Now it would use that weapon as a club, to smash his head in. To beat him and beat him. Have him cry out in agony.

I . . . am . . . not . . . afraid.

The Mothman towered over him. So huge.

I . . . am . . . not . . . afraid.

Manfred gritted his teeth, his entire body trembling.

He's making you afraid. Fight it.

A sound slithered from the creature's mouth, a combination of a purr and a gurgle. Could that be how they laughed?

It lifted the machine gun over its head.

Fight it!

The Mothman swung down.

Manfred brought up his rifle and blocked the blow. Tremors raced up his arm.

The monster hissed. It hefted the machine gun and brought it down again. Manfred knocked it aside, then thrust the rifle's wooden butt into the Mothman's thigh. It let out a bubbling growl and stumbled back.

He took the opportunity to spring to his feet and bring his rifle around. The Mothman clubbed the Gewehr aside. Manfred swung his weapon back at the beast. It jumped back and swept up the machine gun. A bang of metal on wood rocked Manfred's arms and shoulders. The rifle flew from his hands.

"Dammit." He dashed for the fallen weapon just a few meters away. Manfred dropped to a knee, picked up the Gewehr, and twisted around.

The Mothman was right in front of him.

Something large crashed into the monster. An enraged snarl followed.

Moritz!

The Great Dane sank its teeth into the Mothman's black, hairy right leg. It blasted out a blood-curdling screech, jerking its leg back and forth. Moritz would not let go. The creature raised the machine gun to strike its attacker.

Manfred roared and thrust his rifle forward. The bayonet ripped into the Mothman's chest. It stiffened. Moritz let go, barking like mad. The monster staggered back.

Manfred plunged the bayonet into its chest a second time. The red eyes dimmed. The beast let out a weak gurgle and toppled over.

Sucking down a deep breath, Manfred stood over his fallen foe. The eyes had gone dark, and the Mothman did not move.

He rammed the bayonet into its chest again, just to be sure.

Urgent barking erupted from Moritz. Manfred swung around.

Another Mothman landed nearby, facing him.

He brought up his rifle. The Mothman sprang.

Several cracks burst to his right. The creature twitched and dropped to the ground. It hissed and tried to rise.

Manfred put a bullet in its head.

"Herr Rittmeister. Are you all right?"

He turned to see von Döring running toward him, followed by Erhardt and Heldmann. Wisps of smoke floated out of their rifle barrels.

"Ja. Ja, fine." He looked down to see Moritz by his side and scratched the dog's head.

The flares dimmed again. New ones took their place, bathing the base in a white glow.

"Look." Heldmann pointed to the sky. "They're retreating."

Indeed, the remaining Mothmen headed away from the base. Manfred noted their direction. Southwest.

"It looks like a victory for us," said von Döring.

"But a costly one." Manfred frowned, thinking about Ewers and the other poor souls dropped to their deaths by the Mothmen. All to make them more terrified, to sweeten their meal.

Hatred built up in him, such as he never felt before. Not even the Entente earned such disdain from him. Like him, they were professionals, serving their countries.

And they were human.

Manfred looked at the dead Mothmen nearby. What the hell even were these things? Demons? Some new species with an innate hostility toward men?

He did not care. These beasts just wanted to kill them, but not before sucking out all their fear to satisfy their vile appetite.

Manfred thought about the terrified Ewers falling to his death. These monsters could not be allowed to continue living.

"Come." He waved to the other pilots. "We must see how many dead and wounded we have. Then we must shore up our defenses in case they return."

They moved about the base, counting six dead. All from falls of great height. Each body he saw set his fury burning like a great forest fire. Enemies though they were, at least the British and French fought for their national interests. As a soldier, he could understand and respect that.

But these Mothmen. What did they fight for? Did they have a cause, or did they kill because they enjoyed it?

A moan caught his attention. He checked to the left and saw someone hobbling toward him, using his rifle as a crutch. Flares continued to hang in the air, and he soon recognized the other man's face.

"Adam." Manfred hurried over to the pilot, followed by Erhardt, von Döring, and Heldmann.

"Are you all right, sir?" asked Erhardt.

Hans Adam growled. *"Nein.* One of those bastards swooped down to get me. I jumped out of the way just in time, but twisted my ankle."

"Come. We'll get you to Doctor Kretschmer." Manfred slung his rifle over his shoulder, then told Erhardt to help. Supporting the injured Adam between them, they headed off to the hospital tent. Von Döring and Heldmann continued to search the base for any more dead and wounded.

"It looks like you may be confined to bed for a bit," Erhardt said to Adam. "At least you will have some much needed rest."

"To hell with that," spat Adam. "I want to get back in my plane and shoot those evil bastards out of the sky."

"You are not alone in that sentiment," said Manfred.

The hospital tent came into view. It made Manfred wonder how busy Dr. Kretschmer must be. After what he observed of the Mothmen's behavior, it should give more credence to his theory about the creatures feeding on people's fear. He no longer needed convincing of that. But their superiors, like General von Armin, may not be as quick to believe them.

The trio took a few more careful steps when Manfred saw a lump to his left. Not a lump, but the broken form of a human being.

Another tally to the death toll. He sighed, then furrowed his brow. There was something familiar about the figure.

Manfred halted and held his breath. Could it . . .

"You two wait here." He slipped his arm off Adam's shoulder and walked to the body. His steps grew slower the closer he got, not wanting to confirm what he already feared.

Manfred stood over the body and stared at its bruised, bloodied face. Tears blurred his eyes, which remained focused on Dr. Kretschmer.

TWENTY-FIVE

Manfred's eyes burned. It took an effort to keep his lids from closing. He couldn't help but glance at his cot. More than anything, he wanted to sleep.

The responsibilities of command prevented that. His attention shifted from the cot to the men in front of him. The *Jasta* commanders Voss, Von Döring, and Dostler, and the chief of base security, Sergeant Holm. Several burning candles illuminated the gathering.

"I have the final tally of our casualties," Manfred began. "A dozen dead and seven wounded." He read off the names of those killed. Along with Dr. Kretschmer and *Unteroffizier* Ewers, they lost one of their pilots, Lieber, four of Holm's soldiers, two mechanics, two supply personnel, and a mess worker. It took effort to keep any emotion out of his voice as he said each name.

"The monsters also damaged three of our planes," added Voss. "I talked to some of the mechanics. One should be repaired in short order, but the other two are grounded until they get new engines."

Manfred scowled. Between the planes downed from enemy action or storms and this Mothmen attack, *Jagdgeschwader 1's* strength was severely depleted. Lieber's death and Adam's injury did not help matters.

"How many Mothmen did we kill?" asked Dostler.

"Four." Manfred's face stiffened. "But who knows how much those deaths have impacted the Mothmen. Nearly a dozen attacked us tonight. But they could have dozens more, or a hundred more, somewhere out there. They might have launched simultaneous attacks on our other positions around Flanders."

Von Döring drew his head back. "You sound like you are treating them as an opposing army, not a bunch of wild beasts."

"I think it is obvious now that the Mothmen are not simply some previously undiscovered animal. They have intelligence. They knew dropping our men from the air would terrify us." He grimaced at the memory of Ewers falling and smashing into the ground. "They knew that would heighten our fear, giving them more of it to feed on. I also believe the previous attacks were probing missions. Learning our strengths and weaknesses, seeing our weapons in action. If tonight is any indication, they are escalating."

"So they could be done with picking off one plane or a few men at a time," said Dostler. "What we saw tonight could be the start of a full-scale offensive against us."

"Wonderful." Voss snapped a hand up in frustration. "Just what we need with the British pushing into the Ypres Salient."

"I agree." Manfred nodded. "This is not an ideal time to be attacked by another enemy." The skin around his nose wrinkled. *Is there* ever *an ideal time for something like that?*

"Onto our more immediate concerns . . ." He looked at Sergeant Holm. "Where do we stand with protecting this base?"

"I've pulled back some of the men assigned to the perimeter and put them around our more vital areas. The planes, the armory, supply tents. Since the Mothmen can fly, we can't rely on our usual defensive positions to stop them."

"Good. But I do not intend to wait for the Mothmen to attack us again. We need to switch to more aggressive tactics to find them."

"What do you propose, *Herr Rittmeister?*" asked Dostler.

Manfred spread a map over his desk. The other men leaned forward as he used a finger to trace a line from their base toward the lefthand corner. "The surviving Mothmen were flying southwest when they retreated from here. I've been studying their possible route and trying to determine where their base might be. If the Mothmen have a sizeable contingent, they will not be operating from a church bell tower. Probably not from any building." He gave a slight shrug. "Of course, there are not many large buildings still standing in this part of Flanders for the Mothmen to use."

He tapped a finger on the part of the map that showed Ypres. "I also cannot believe they would stage attacks out of the Salient. There is too much fighting there, too many people. They could be easily discovered and destroyed. I believe they would want to establish a base away from people, a place that offers good concealment."

Manfred slid his finger up from Ypres, past a river, past what was labeled an abandoned trenchline, and to a puffy mass of black.

"This forest, just west of Meulebeke."

"Why that forest in particular?" asked Voss.

"It is along their projected flight path from our base," said Manfred. "When you look at the radius of their attacks, this forest is almost in the middle of it."

Von Döring nodded. "Which means less time in the air for them, less time to be spotted and attacked. And our reconnaissance planes would never be able to find them among all those trees."

Dostler exhaled, hands on his hips as he gazed at the map. "So now that we . . . well, *think* we know where they are, how do we attack? Randomly strafing parts of the forest would be a waste of bullets. Going in there on foot would invite the Mothmen to ambush us."

"You are correct." Manfred glanced at the commander of *Jasta* 6 before returning his attention to the map. "So instead, we will flush them out of the woods."

"How?" asked Voss.

"We can task a group of bombers to drop their loads onto the forest. Hopefully, whatever Mothmen are not killed will evacuate their hiding places. And we shall be circling above, ready to pounce when they reveal themselves."

"You said bombers." Voss held up a finger. "General von Armin will not even give us a squad of soldiers to help defend our base. Do you think he will divert even one bomber from the Ypres Salient for this operation?"

Manfred straightened. "I will make certain of it. Because tomorrow, I am flying to Fourth Army Headquarters to personally make my request to the general, and I do not plan on leaving until he agrees to it."

TWENTY-SIX

Manfred held his breath as the roof of the stone manor just outside the city of Ghent appeared on the horizon. Reputation or not, he was putting his career at risk by walking into Fourth Army Headquarters and demanding to speak to General von Armin without making a proper request. He was sure to be reprimanded for such a breach of protocol. Right now, he did not care.

And if the general is as serious about stopping the Mothmen as he claims, he will grant my request.

At least, Manfred hoped that would be the case.

Within minutes, he was able to make out the details of the red brick manor. The ornate home was three stories high with rounded towers on either side, each capped by a pointed roof. The expansive lawn led up to a fountain with what Manfred thought were two leaping porpoises.

Then there were features the builders of this manor could never have envisioned. Staff cars, trucks, and motorcycles parked near what appeared to be the stables. Sandbagged machine gun nests along the perimeter. And his biggest worry, the four Flak L/30 anti-aircraft guns set up at each point of the compass. Having given no prior announcement of his arrival, the gunners might mistake him for an enemy and open fire.

One lone plane far from the front, they will assume it is German.

Still, he took no chances. He turned the Albatros D.V. to the right and dipped the wing, wanting to give any soldier with binoculars a clear view of the black iron cross on the side of the airplane.

Manfred circled the manor. His muscles unwound when no one fired at him. Finding an open patch of grass half-a-kilometer from the home-turned-military headquarters, he set down the plane. The propellors had just started to slow when the breakfast he had quickly eaten before taking off for Ghent rebelled. Bile surged through his gut and up his throat. He swallowed hard to push it down. A crushing pain engulfed his head. He leaned forward, closing his eyes and resting his head against the instrument panel. He'd wait until the discomfort settled before making for the manor. Manfred wanted a clear head before meeting General von Armin.

A growling noise caught his attention. It was one he'd become familiar with recently. A motorcycle. For a second, he thought Voss had followed him here on his damned machine. But that couldn't be.

He opened his eyes and looked to his left. A motorcycle with a sidecar rumbled toward him. The driver halted a few meters from his plane and stared at him through goggled eyes.

"Herr Rittmeister von Richthofen?"

"Yes."

"Ah. We thought that was your plane." The young man dismounted. "My lieutenant sent me to collect you. *Unteroffizier* Braun at your service." He saluted.

Manfred returned the salute. Not only had his fame saved him from being shot by his own side, it earned him a ride to headquarters.

He climbed out of the cockpit and into the sidecar. Braun hopped onto his seat, started the engine, and sped toward the manor. A smile crept across Manfred's lips when he set eyes on it, thinking of Dr. Kretschmer's comment about generals running the war from mansions far removed from the frontlines.

That smile faded when he thought back to the staff surgeon's crumpled body lying on the ground.

Braun parked near the wide, stone stairs leading to the front entrance. He escorted him to General von Armin's office. There, a pretty secretary with dark hair looked up at Manfred with wide eyes.

"Herr . . . Herr Rittmeister von Richthofen," she said with breathless surprise.

Manfred paused for a moment, admiring the secretary's smooth face. In these times, one should admire a beautiful woman when the opportunity arose.

Unfortunately, his duty prevented him from enjoying that pleasure for long. "Please tell General von Armin I am here to see him."

"Of course, but I do not recall your being on his schedule for today." She started to shuffle papers.

"Because I am not," he stated. "This is an urgent matter."

"The general is conferring with his staff. He may be some time."

"Tell him I am here with critical information on the . . . special mission he assigned my unit. He will understand."

The secretary sucked on her lower lip for a second. *"Ja, Herr Rittmeister."*

She rose from her seat, her slim body tense. She walked up to the door, hesitated, then knocked.

"Yes?" came the muffled bark from General von Armin.

The secretary opened the door and drew a quick breath. "Apologies for the interruption, sir. But *Rittmeister* von Richthofen is here to see you."

"What? I did not send for him." The agitation in von Armin's voice was evident. "Why is he here?"

"He says he has critical information about a special mission you assigned him."

Manfred's shoulders knotted as he waited for the general's response. Would it be anger? Would he tell him to go back to base and submit his request through proper channels as was expected for an officer in the Imperial German Army?

"Inform the *Rittmeister* I will see him in twenty minutes."

Manfred's muscles loosened as the secretary replied, *"Ja, Herr General."*

He took a seat across from the young woman's desk. She fetched him coffee, which he drank while he waited.

Precisely twenty minutes later, a parade of officers filed out of the office, ranging in rank from *hauptmann* – captain – to *oberst* – colonel. He stood at attention as they passed. Many eyed him with surprise. One, a major, glared at him for some reason. Perhaps jealousy. Maybe Manfred's presence had ended the meeting before the man could present some no doubt brilliant idea to the general that would inch him closer to a promotion.

Headquarters types.

"Rittmeister."

Manfred turned to see von Armin standing in the doorway, his face pinched in obvious anger.

"In here."

Posture erect, he walked into the office. Von Armin shut the door, hard.

"This is highly inappropriate, *Rittmeister.*" The general stalked over to his desk. "I have a major battle to oversee." He jabbed a hand at the table next to Manfred, which had maps of the Ypres Salient laid out.

"The British have seized part of Westhoek," von Armin continued. "They almost took Inverness Copse and Glencorse Wood, but our forces recaptured them. Now we have reports of British and French troops massing from Bixschoote to Langemark. So as you can see, I have little time for an *unscheduled* meeting with a junior officer, regardless of his fame and accomplishments."

"This does have to do with the Mothmen, sir."

"And you could not send a report about it, or make an official request for a meeting?"

"They attacked my base last night," declared Manfred.

Von Armin went still. The lines that had dug into his face faded. To Manfred, it seemed the seriousness of his statement had sunk in.

"How bad was the damage?" asked the general.

"Twelve dead, including our staff surgeon, Dr. Kretschmer. I also had one pilot killed and another injured, with three planes damaged."

Von Armin grimaced at the news. "How many of these monsters attacked your base?"

"At least a dozen."

"A dozen?" von Armin's head jerked in surprise. "I thought there were just a handful of these Mothmen."

"Apparently not, sir," said Manfred. "I suspect there are even more. I also thought what happened last night could be a coordinated attack on us by the Mothmen. Have there been reports of them attacking any of our other positions in Flanders?"

"No." Von Armin shook his head. "We've had no such reports come to headquarters."

Manfred took a couple of steps toward the desk. "I believe what happened last night could be the start of a campaign by the Mothmen."

The general grunted. "It could not come at a worse time as we are fighting all along the Ypres Salient. Have you found where these monsters are coming from?"

"I think so, sir." Manfred removed a rolled-up map from the tubular case attached to his belt. He unfolded it on von Armin's desk.

"When they retreated from our base, they headed southwest. Now if the Mothmen do have large numbers, they likely will not dwell in buildings. Especially as there are not many large buildings still intact in this part of Flanders."

"They are probably sheltering with their British or French masters."

"I do not think the Mothmen are working for the British, or the French, or anyone else." He stared von Armin in the eyes. "I watched them up close last night, even looked one of them in its red eyes. They are . . . utterly inhuman. I think they look upon all men the same, regardless of uniform. We exist for them to torment, to feed off our fears, and then to kill off brutally."

Von Armin said nothing.

Manfred's lips tightened. Did the general believe him? He had no conclusive proof. Just theories. Perhaps if Dr. Kretschmer was still alive, he could come up with some scientific explanation to back up his speculation.

The silence lasted a few more seconds, until von Armin took a breath and said, "So you believe they are operating inside our lines?"

Manfred stifled a sigh of relief. He wondered if the general would have believed this from any other junior officer who lacked the name von Richthofen.

"I do, and judging by their line of retreat, I believe the Mothmen are hiding here." Manfred stabbed the map with his index finger. "These woods near Meulebeke."

Von Armin bent over a bit, focusing on that spot on the map. "Are you sure, *Rittmeister?*"

He chewed on his lip for a moment. "It is my best guess, sir."

"But you have no proof of this." Von Armin frowned. "Send reconnaissance flights to the area."

"That is the reason they would use the woods as their base. Our scouts would not be able to see them through the trees. Not only is it a perfect hiding place for the Mothmen, it is almost in the center of the radius of their previous attacks. That would mean short flight times to their targets, reducing their chances of being spotted."

Von Armin straightened. He pinched his chin between his thumb and index finger, staring in thought at the map. "If what you say is true, how do you propose we deal with them?"

"We flush them out," said Manfred. "I can have my squadrons circle the area while bombers drop their ordnance on them. We are certain to kill some of them. The others, they will realize we have found their location and will retreat. That's when my squadrons will dive down and eliminate them."

Von Armin shook his head. "How can I spare any bombers with all the fighting along the salient?"

"Then artillery." Manfred noted the hint of desperation in his tone. "Just two or three batteries should suffice."

"And you think I can spare any artillery? I need every gun available to hold back the British and French."

"Sir." Manfred flattened his palms on the map. "If the Mothmen attack our positions like I believe they will, it could disrupt our entire front line, allowing the Entente to break through. They could advance across Belgium, and that would put hundreds of thousands of British and French soldiers on our border."

Von Armin winced at the comment.

Manfred continued, "By that time, the Americans could be here in force. Then all three armies could invade Germany itself."

The general hung his head. His shoulders rose in a long breath. Several seconds passed before he looked back up. "You are sure this is where the Mothmen are hiding?"

Manfred stood straight. "Yes, sir. I will stake my reputation on it."

Von Armin regarded him silently for a few seconds. "Very well. I will shift three artillery batteries to bombard the woods near Meulebeke to drive the Mothmen from it and into the guns of your pilots."

"Thank you, sir."

"You had better be right about this, *Rittmeister.*" Von Armin's eyes narrowed. "Because if you are wrong and those batteries are out of position at a crucial time, then it might be you and not the Mothmen who cost us this battle. And, perhaps, the entire war."

TWENTY-SEVEN

Manfred checked his watch as strong winds rushed over his open cockpit. Five minutes until the artillery rained down on the woods outside Meulebeke. He looked at the landscape below, the green mass of the trees, the river to the west, including the bridge guarded by the mobile anti-aircraft gun. A few kilometers beyond that lay the abandoned trenchline. He frowned, wondering if it would be used again should the Entente advance deeper into the salient.

His gaze shifted from the sky to the woods, just in case any Mothmen decided to fly up and attack them. None did, but Manfred was ready for them. His *Jasta* 11, along with von Döring's *Jasta* 4, circled what remained of the town of Meulebeke. Flying directly over the woods would put them at risk of being hit by incoming shells. Two thousand meters higher, Voss's *Jasta* 10 covered them. In every engagement so far, the Mothmen had the element of surprise on their side. No more. Now they would be the ones to be surprised.

So long as they are in those woods.

Manfred scowled at the thought. He tried to push away his doubt. The woods were the most logical place for the Mothmen to hide.

And what if these creatures do not think logically? What if they are far away from here?

His shoulders tensed. Worrying was pointless. His plan could not be changed now. It would either succeed or fail.

And if it fails . . .

A puff of gray below caught his attention. A plume of smoke rose from the woods. More soon sprouted.

Manfred breathed in halfway and held it. *It has begun.*

Explosions erupted along the entire length of the woods. It did not take long for grayish-black clouds to form and blot out many of the treetops. Manfred observed the edge of the woods, waiting for the dark shapes of the Mothmen to flee the bombardment.

None did.

Perhaps they had taken cover until the shelling stopped. But they would eventually flee. Manfred had no idea how many Mothmen were in Flanders. He could not imagine them having numbers on par with any of the armies in the field. Otherwise, they would have seized large swaths of land and established bases throughout Flanders.

No, they wished to remain hidden, coming out to attack only when it suited them. With their encampment discovered, they would have to abandon it and find another.

Come out, damn you.

More explosions tore through the woods. No flying monsters emerged from the trees.

Manfred's eyes scrunched behind his goggles. Something was different about the smoke hovering over the battered woods. Mingled among the gray and black were streaks of yellow.

Mustard gas.

He sat up straighter in his seat. The Mothmen might be able to shelter in a hole or behind trees until the shelling ceased. But gas, that was another matter. He could not imagine the monsters possessed gas masks. Unless they wanted their lungs seared, they had to leave the woods. Now.

Manfred's gaze swept up and down the treeline, and the smoke above it. Anytime he should spot the Mothmen. Anytime.

Not a single creature appeared.

He clenched his teeth. *Where the hell are you?*

Manfred snorted. If the Mothmen did not evacuate their positions, his *jastas* would have to land in a field about three kilometers from Meulebeke and go into the woods on foot. It was not an idea he relished. The monsters could easily hide in the treetops and dive on them. Also, he'd have fewer than twenty men to search the woods. A couple of infantry companies would have been ideal for the task, but he had been fortunate to get General von Armin to commit the artillery batteries for this operation. He doubted the chief of 4th Army would part with three hundred or so soldiers while fighting raged across the Ypres Salient.

The bombardment ended. Manfred and his planes continued to circle the wrecked town for five minutes. Ten minutes. No Mothmen appeared.

Perhaps they are tending to their wounded before they leave.

He waited five more minutes. No monsters flew out of the trees or smoke. Manfred grimaced. They had no choice now. They would have to search the woods for the Mothmen.

His gut tightened at the thought of finding no trace of the creatures. He had staked his reputation on them being in these woods, convinced General von Armin to divert resources from a major battle. If he was wrong, he doubted all his past accomplishments would save him from the consequences. The general would probably reassign him to supply or some other mundane task that would drive him mad.

He looked at the colorful planes around him, ready to give the signal for them to land and begin their trek toward the woods.

Manfred jerked when a plane to his right burst into flames. It slewed away from him and spiraled to the ground.

He twisted around in the cockpit and looked up.

Eight Sopwith Camels dove on his squadrons.

TWENTY-EIGHT

Manfred shoved the controls left. The Albatros banked hard. He glimpsed the other aircraft of his *jasta* doing the same. All following one of the tenets of *Dicta Boelcke*. When the enemy attacks from behind, do not run away. Turn and face them.

The turn completed, Manfred yanked the stick back. The Albatros rose just as the first two Camels zipped past, still in their dives. He thumbed the fire buttons as the remaining six soared by. He had no idea if he hit any of them.

Gritting his teeth, Manfred threw the Albatros into a sharp turn. Cold wind screamed around him. A crushing weight pushed against his body. Pain squeezed his skull. For a second, he feared his brain would explode.

He grunted, fighting through the agony. *Press the attack. Press the attack.*

The British appeared below him, pulling out of their dive. Manfred checked around him. The rest of his pilots swung around to join the attack.

His eyes darted from one Sopwith Camel to the next. The newest fighter in Britain's Royal Flying Corps, it was more maneuverable than his Albatros D.V. But positioned above them, he had the advantage.

Manfred picked one Camel to his right. He pushed the Albatros' controls forward, putting the plane into a shallow dive. He glanced between the enemy fighter and his gunsights, predicting where his prey would be seconds from now. His thumbs hovered over the firing buttons.

Wait . . . Wait . . . Now!

The two LMG 08s chattered. The Camel passed through his gunsights. The aircraft shuddered. A long tongue of flame stretched from its side. The Camel rolled over and fell from the sky.

Manfred nodded. Kill number fifty-eight.

But he had no time to celebrate. Seven other Camels remained.

Six, actually. He spotted Heldmann's Pfalz D.III riddle another British plane with bullets. Streaming smoke, it plummeted toward earth.

So did one of Manfred's planes. Another Camel flew head on toward Plass' fighter. Both pilots blazed away with their machine guns.

The two aircraft shot past each other, separated by no more than ten meters. The British pilot flew on. Fire and smoke gushed from Plass' Albatros. Manfred grimaced as he watched his pup go down.

He would have to mourn later. He swung his head in all directions, seeking out another target . . . or another threat. Two more of his pilots were being chased by Camels, including Gude. Manfred swung his Albatros left to help.

Smoke suddenly spewed from the Camel behind Gude. Manfred craned his neck.

Voss and the rest of *Jasta* 10 dove into the fray.

The odds had become stacked against the British. Still, they showed no signs of retreat. He commended them for their courage. Outnumbered though they were, they were determined to shoot down as many planes as possible. Manfred and his men would probably do the same in their position, especially given the desperation of the battle along the Ypres Salient.

His eyes darted about the sky until he spotted a Camel below and to his right. It flew away from the fight, either disengaging or trying to come around for another attack.

Whatever the case, it would be kill number fifty-nine.

Manfred turned, keeping above the British fighter. Waiting for the right moment to dive and –

A dark form flashed between the two planes. Manfred's chest clenched.

A Mothman.

It swung toward the Camel and flapped its wings twice. In seconds it was just above the fighter's rear. The monster's hand lashed out. The tail shattered into a storm of splinters.

Manfred's eyes narrowed. The Mothman had confirmed his suspicions. These creatures did not serve the British or French. All men, regardless of uniform, were their prey.

The Camel wobbled. With a flap of its wings, the Mothman bounded over the fuselage. It attached itself next to the cockpit as the plane tilted right.

Manfred scowled. A damaged plane, a terrifying monster barely inches away, the British pilot had to be scared out of his mind. And that damned abomination was feeding on that fear.

He swung toward the Camel, its nose dropping toward the ground. The Mothman still clung to the side. With the British flier taking no evasive maneuvers, it made it easy to line up for a shot. Still, Manfred hesitated, a touch of sympathy surfacing for the enemy pilot. Right now the Mothman was his primary target. The Englishman just happened to be in the way. Or more likely paralyzed by fear. Not the way he wished to bring down another pilot. There would be no honor in this kill. But the man in the other aircraft was doomed anyway. Either the monster would kill him with its claws or just let him crash into the ground when it had filled its – belly? Mind? – with his victim's terror.

Manfred thumbed the fire buttons. Machine gun rounds chewed through the Camel. Wood splinters jumped off the fuselage. Smoke streamed from the plane. The slack form of the Mothman dropped away from the fighter and tumbled toward earth. The Camel soon followed.

He searched the sky around him. Everywhere he looked planes twisted, climbed, or dived. Dark, winged forms pursued them. Or in some cases, fled from the aircraft. Manfred struggled to count the number of creatures they faced. A dozen, perhaps?

He caught sight of Heldmann firing a burst at one of the monsters, missing. A Camel also fired at a Mothman. It seemed at least one of the Englishmen realized they now had a common enemy.

An Albatros D.III plummeted toward earth. One of the monsters pushed off its side, rolled, and flew off in search of other prey. To Manfred's right, a Mothman went into spasms as Voss riddled it with his machine guns.

Manfred checked left. His chest tightened at the sight of a flying monster chasing Erhardt's black and orange Albatros D.III. The *leutnant* snapped his fighter into a tight turn, trying to bring the fight to the Mothman. But smaller and lighter than the plane, and flying naturally instead of by machinery, the creature banked quicker.

He jerked the controls left. He had to help his wingman. But as the gap closed between the Mothman and Erhardt's fighter, Manfred knew he'd be too late.

Erhardt pushed the D.III's nose down. The plane went into a dive just as the monster soared past, missing it by only a few meters. Manfred let out a quick sigh of relief.

The Mothman looped over and again went for Erhardt. But the *leutnant's* maneuver gave Manfred the time he needed to get into position to aid his wingman.

The Mothman leveled out, its body aimed straight at Erhardt's D.III as it pulled out of its dive. Manfred nodded. The beast was doing something he had preached to his pups to never do. Fixate on their target.

Manfred drifted left, leading the Mothman. He waited two beats, then fired his machine guns. Barely a second later, the creature crossed his gunsights. It twitched, went limp, and fell.

Exhaling, Manfred nodded. Two Camels and two Mothmen. A very good day of hunting.

The day is not over.

He turned the Albatros left to catch up to Erhardt.

Something jolted the plane. Manfred jerked in his seat, the straps the only thing keeping him from being thrown out of the cockpit. He cried out as pain hammered his skull. Crushing the controls, he fought to right his fighter. But it heeled to the left. As if some weight was clinging to the –

He snapped his head left.

Right in front of him were the blazing red eyes of a Mothman.

TWENTY-NINE

Manfred's lungs froze. So did the rest of his body. The growl of the engine, the wail of the wind. It all faded into the background. His world condensed to the pair of red orbs.

Fear flooded his being, like water exploding from a burst dam. This . . . inhuman thing was right in front of him. Staring at him . . . through him. Would it slash his chest open? Just let him crash and die in flames?

Not like this. Please. Don't kill me.

The plane dipped farther to the left. The Mothman swayed back, as if it lost its grip. It raised its right hand and slammed it into the wooden frame behind the cockpit. The plane rocked for a second. Pain squeezed Manfred's head. For a moment, he feared his brain would burst through his skull.

Through the agony came a sliver of realization. His terror had fallen away. But only briefly. He felt it rise again, ready to consume him.

No.

The fear came at him like a rushing wave.

No! This is not you.

Cold needles pricked the flesh under his flight jacket. He could not move.

It's all in your mind.

He refocused on the Mothman's eyes. Such a bright, burning red. Igniting all his fears like a wildfire.

Yes. It is making you afraid.

Groaning, Manfred slowly raised his shoulders.

I . . . am . . . not . . . afraid of you.

Air hissed between Manfred's teeth like escaping steam. Fear was nothing new to him. It was the constant companion of every soldier. But the good ones knew how to suppress it. How to keep it from crippling them in the most dire of moments.

And Manfred von Richthofen was a good soldier!

He gazed directly into the Mothman's red eyes. *I am not afraid of you.*

Manfred yanked out his Luger.

The Mothman caught his movement. It tore its right hand from the fuselage.

Too late. Manfred jammed the barrel against its forehead and pulled the trigger. Red gushed from the back of the Mothman's head. It slipped off the side of the Albatros and spun away . . .

And smashed into the tail section

"Dammit." Manfred spat as the left side elevator shattered in a storm of splinters.

He turned forward. His eyes went wide when he saw the ground rushing up. His altitude couldn't be more than 800 meters.

Manfred pulled back on the controls. The Albatros bucked. With one of his elevators gone, controlling the plane's pitch would be difficult.

Then he saw the upper left wing. A huge chunk had been torn from it. The Mothman's doing, no doubt.

 He wrestled with the controls, trying to keep the nose up. At the same time, he reduced his speed. He was going to land, and land hard. The key was to not land so hard the Albatros exploded and scattered his remains across Flanders.

Manfred's arms strained to keep the fighter steady. But whenever he twisted the control one way, the aircraft pulled in the opposite direction.

A copse of trees appeared before him.

Teeth bared, sweat drenching his skin, he got the nose up, up . . .

Then the Albatros slewed right.

He yanked the controls toward his chest, letting out a primal howl as he did. The Albatros rose, barely. The trees were so close he could touch their tops.

Then they were gone. A field stretched out in front of him . . . then turned into a group of small hills less than a kilometer away.

If he was going to land, he had to do it now.

Again he cut his speed, struggling to keep the nose level. The plane wanted to drop to the left. The muscles in his arms burned as he fought to keep control. The ground got closer. Forty meters . . . Thirty meters . . . twenty. He was going to . . .

Half of the top left wing ripped away from the plane and spiraled out of sight.

The Albatros slewed right. Manfred sensed he was about to flip over. With another howl, he twisted the controls left.

A sharp blow hammered the plane. Manfred's world shook.

THIRTY

Manfred blinked. The world came back into focus.

I'm alive. That's good.

Alive, but with an invisible force crushing his skull. He leaned forward in his seat, grasping the sides of his head, willing the pain to cease.

It did not.

Manfred grunted and closed his eyes. *Stop . . . Stop,* he begged whatever threatened to squeeze his brain until burst like a rotten orange.

You are in enemy territory, whispered the warrior part of his brain. *Stay aware.*

He inhaled and leaned back in his seat. Manfred breathed deep again, and again. Technically, he was in German territory. Though a part apparently controlled by the Mothmen. He could not afford to succumb to his ailments.

Manfred stopped focusing on his pounding headache. That had, unfortunately, become a normal part of his life. Did anything else hurt?

His shoulders throbbed, probably from the straps digging into him when he crashed. If not for them, he would have been thrown into his instrument panel, his skull shattering. Dying the same way as his mentor, Oswald Boelcke.

He undid his safety harness and rotated his shoulders. Twinges of pain made him grimace. Sore, but it did not appear he'd dislocated anything. Manfred next moved his legs and feet. They, too, were sore, but did not feel broken.

Manfred was in better shape than his plane. It rested on its left side, the lower wing gone. The left landing strut must have also snapped off when he hit the ground. The Lord must have been watching over him to come away from this crash without serious injury.

Groaning, he pushed himself out of the cockpit and slid to the ground. His legs went rubbery, forcing him to grip the side of his plane. Then came the nausea, roiling through his stomach. Manfred did not try to stop it. He dropped to all fours and vomited.

Spitting out the last of the stale, hot bile, he rose, took a couple of deep breaths, and walked in a small circle next to the wrecked fighter. He winced from the pain shooting up across his ankles and legs. While it hurt, he could still walk, confirming that nothing was broken.

Mothmen, his brain warned him.

He went for his holster, but his pistol was gone. Had he dropped it when the Mothman hit his elevator? Instead, Manfred reached for his knife and gazed at the sky. Planes still twisted and dove above him. From his vantage point, he could not make out any flying monsters. But he had no doubt they were still battling his *jastas* . . . and likely the remaining British Camels.

One fighter tumbled out of the sky. Manfred had no idea which side it belonged to. His shoulders sagged as he held the plane in his gaze, watching it fall closer to earth, and its ultimate doom.

He tensed when the thought hit him. *What if a Mothman saw me crash?* The thing could be descending to check out the wreckage for a survivor . . . and finish him off. Likely after feasting on his victim's fear.

They will be sorely disappointed if they try to do that to me again. After what just happened, he was confident he could resist the Mothmen's . . . mind power the next time he fought them.

Manfred hurried to the cockpit, grabbed his rifle, and shoved in the five-round stripper clip. He then got his cavalry sword and sheath and strapped it around his waist. The Luger, however, was nowhere to be found. It must have slipped from his grasp when the monster struck his tail section. He'd been so focused on controlling his damaged plane that he hadn't even realized he'd lost the pistol.

Fixing the bayonet to his rifle, he set off, taking one last look at his fighter. All it would be good for now was salvaging for spare parts. He'd need a new plane. That suited him fine. He never cared much for the Albatros D.V. It had poor maneuverability and the heavy controls made it physically taxing to fly. The time had come to get a better aircraft.

Perhaps the triplane Fokker had me fly.

Manfred shook his head, clearing his mind of the ruminations. He needed to concentrate on fighting and surviving.

He set off across the field, eyeing the hills nearby. He needed to get atop one of them and scout his surroundings before determining his next move.

Several minutes later, he started up the first hill he came upon. His legs grew heavy walking up the steep angle, but he pressed on. Soon he reached the top and lay flat on his stomach. Not only to rest, but to make himself less of a target. Standing on a hilltop would invite any Mothman in the area to come after him.

Manfred removed the binoculars from his pouch. The woods the artillery had shelled lay about two-and-a-half kilometers to the northeast. Before that was the abandoned trenchline. He lowered the binoculars and exhaled slowly, thinking.

The smartest course of action would be to head south to the river, then hike along it until he reached the bridge guarded by the anti-aircraft truck and the infantry platoon.

But then he'd be neglecting his duty. To find the Mothmen's nest and destroy it. As one man, he wouldn't be able to carry out the latter. But the former was possible. The monsters had to be somewhere in the woods. If he could pinpoint their nest, he could bring to bear more firepower to finish them off.

Unless they decided to abandon their nest. Though considering the attack on his *jastas* and the British squadron, perhaps the Mothmen decided to hold their ground.

Whatever the case, he needed to know for certain.

Manfred got to his feet and descended the hill, much less strenuous than going up. He trekked across the field. All around him was grass, marred by an occasional shell crater. Probably from the last major battle for the Ypres Salient two years ago. Other than that, he had nothing to use for cover.

Manfred's neck and shoulders tensed. He looked up every few seconds. Any Mothman would be able to spot him with ease. But the fighter planes still zipped and banked above. It appeared the creatures would be busy with them.

For now. His pilots would soon be low on fuel and have to withdraw. He imagined the Mothmen would be tired, too. They would probably return to their nest, either to dig in for another attack or retreat. Whatever the case, the monsters would lead him to it.

Manfred hiked across the field, taking shelter in any crater he came across. He checked his surroundings, making sure no Mothmen were near, took a swig from his canteen, then moved on.

Little more than a kilometer from the trenchlines, he came across another crater and eased himself into it. Like the other ones, the recent rain had left it muddy. He could not take the chance of slipping and burying his rifle in the damp earth. He could not afford to have the weapon jam if he came across any Mothmen.

Manfred checked across the grassy expanse. Clear. He looked up.

Three of his planes descended. He easily identified their pilots from the color schemes and types. Heldmann, Gude, and Erhardt, who trailed the other two by a wide margin. Manfred soon understood the reason for the sizable gap. A pair of dark, moth-like wings appeared between Erhardt and the other two fighters.

Erhardt's DIII lifted its nose. Specks of orange flashed on the cowling. The monster tumbled from the sky.

Manfred watched the three planes fly closer to the ground. He guessed they would land in the field outside Meulebeke, just as they had discussed back at the air base, and try to locate the Mothmen nest on foot. He decided to make his way there and link up with his pups. His chances for success, and survival, were better operating as a team than doing this alone.

He headed east, quickening his pace. Manfred would never reach them by the time they landed. If he was fortunate, he could join up with them before they headed into the woods.

Another check around him showed no one else in the area. Manfred then stared at the sky . . . and halted.

Several planes turned north. His *jastas*. Three other planes made for the south. They had to be what remained of the British squadron. As he expected. The fighters would have to disengage before their fuel tanks ran dry.

What Manfred could not be sure of was whether or not the Mothmen would return to their nest or pursue the planes.

Best to err on the side of caution. He would assume the creatures had their fill of his men and the British and come home.

Which presented Manfred with another dilemma. If their route to the woods passed over his position, the Mothmen would likely spot him.

Exhaling, he looked around, glimpsing a single plane flying east. He had no idea if it was German or British, or why it chose that direction. Pursed by Mothmen, perhaps?

He returned his attention to the field. There was nothing to offer him full concealment from the monsters. The abandoned trenches, by his estimate, lay about three-quarters of a kilometer away. Perhaps he could find some alcove or dugout to hide in. Provided he could reach it before the Mothmen began their descent. Of that he had no guarantee.

He banged a fist on the lip of the crater. *Think.* Again, Manfred studied the field, for all the good it did. Nothing was close by to shield him from the airborne eyes of the Mothmen. All he could do was lie in this muddy hole and pray the beasts did not notice him.

That's when he looked at the bottom of the crater. Manfred pressed his hand into the dark earth. It was not a soupy brown mess, but thick, damp. And it gave him an idea.

He removed his flight jacket, scooped up handfuls of mud, and rubbed it on his uniform. He then smeared more of it on his face. Wrapping his jacket around his rifle, he laid it against the side of the crater. The best he could do to keep any mud from getting into the barrel and the chamber. Plus his jacket was dark brown, and matched the color of the ground . . . or close to it.

Manfred pressed himself against the dirt next to his rifle, peering over the edge, and waited, trying not to move a muscle.

Several minutes passed, he guessed. He dared not look at his watch. But the sky and ground remained devoid of flying monsters. Would they pass this way? How long should he wait? Instinct and training told him to stay here until night and move under cover of darkness. That he could do were he up against the British or French. But from what he'd seen, the Mothmen operated just as well at night as in the daylight. He doubted the veil of black would keep him safe from those blazing red eyes.

Another worry surfaced. Not for him, but his three pilots who landed near Meulebeke. Had the Mothmen spotted Heldmann, Gude, and Erhardt? Did they not take into consideration the threat from above? Would they not take proper precautions?

Those three are not raw recruits. They are good, experienced men. They will know what to do.

The thought, however, did not ease his worries one bit.

Manfred lay in the crater. A gentle breeze broke the stillness. Perhaps the creatures had kept after the planes. If so, he could risk making for the trenches and the woods beyond. He just had to figure out how long he should wait before . . .

A shadow drifted over the grass to his right.

Manfred held his breath and went absolutely still.

A dark form appeared in his peripheral vision. He resisted his fighter pilot's urge to turn his head in the direction of a potential threat. Seconds later, a Mothman glided above the field. Then came another. Another. Soon more than a dozen creatures flew overhead.

Manfred drew a short, quiet breath. He still feared it was loud enough for the monsters to hear. How good was their hearing anyway? Could they pick up the rapid pounding of his heart? The way it echoed in his ears the British could probably hear it on their side of the salient. Or maybe his muddy camouflage was not as good as he thought. Maybe one of them would see him, whisk him off the ground and drop him like they did with poor Ewers.

He shut his eyes tight. *I will not be afraid. They will not suck out one ounce of fear from me. I will not give the bastards the satisfaction.*

None of the Mothmen saw him. They continued flying. Manfred tracked them, wanting to see where they would enter the woods.

But they never reached the woods. Every single Mothman descended into the abandoned trenches.

THIRTY-ONE

Why there? Manfred stared at the trenches as the last few Mothmen dropped behind the mounds of dirt and weathered sandbags. Perhaps they wanted better cover in case of another barrage.

Or they could just fly away. Again, he wondered if they had every intention of defending their base, or nest, or whatever they called it. Could one patch of land be so important to creatures that could fly?

That's when his eyes widened in realization. *Nest.* Any bird would defend it to protect their young if attacked by another animal. Could that be it? Could the Mothmen have young nearby? Young that may not be able to fly?

He swallowed, dreading the thought of more of these monsters soaring around Flanders, and producing more babies. Then those creatures spawning more of their kind and on and on. Manfred's stomach twisted at the vision of an army of Mothmen blotting out the skies of Europe, drinking in the terror of millions of men, women, and children. Then slaughtering them with as much thought as a man gives to stepping on a cockroach.

His face went stiff. *I will not let that happen.*

Manfred waited roughly five minutes. No more Mothmen appeared. Holding his breath, he dared to turn his head and look into the sky.

All was clear. No flying monsters or airplanes overhead.

He then observed the trenches for several minutes. None of the creatures poked their heads over the earthworks or sandbags. Manfred ground his teeth together, thinking. The prudent course of action would be to get away from here, report the Mothmen's position to the Army, and have them hammer the trenches with artillery.

But by the time I reach any unit, they could be gone. Even if the monsters remained in the trenches, there was no guarantee the bombardment would kill them all. How many millions of shells had his army rained down on British and French trenchworks over the past three years and they hadn't killed every single enemy soldier.

He needed more intelligence. Were they using the trenches for a brief respite or did the Mothmen live there? If the latter, where in the trenches did they live? Had they modified their surroundings? Did they have any supplies? Food? They had to eat actual food, not simply live off the fear of their victims. Didn't they?

Manfred gazed out at the open field, then to the sky. This is something that should be done at night. But as he'd thought before, would that make a difference to creatures like these?

Exhaling, he unwrapped his flight jacket from around his rifle. With one final look around and above – *all clear* – he hauled himself out of the crater. Flat on the ground, he crawled through the grass and dirt. Fear skittered through his mind like a horde of ants. Not the doing of the Mothmen, but his own. Fear that one of the damned beasts would peek over a sandbag and see him.

Manfred scowled at his rifle and the five bullets within. How he missed his plane with its two machine guns.

But he kept crawling, his heart pumping harder the closer he got to the trenches. Sixty meters . . . fifty meters. His eyes shifted back and forth. Any second he expected to see a Mothman, who would sound the alarm. Then he'd have to face more than a dozen monsters all by himself.

Perhaps this was not a good idea.

Twenty meters . . . Ten meters. Almost there. He prayed for the creatures to stay concealed.

Manfred got to the edge of an earthen berm. His muscles loosened in relief, but only for a moment. Somewhere within these trenches lurked the Mothmen.

He crept to the tip of the berm, froze for a second, then peeked over the top as much as he dared.

The trenches were empty.

He crawled farther along, peering down from the piles of dirt. Still no sign of any Mothmen. Manfred crept another ten meters or so before he came to a curve in the trench line. He clutched his rifle tighter. Should he risk it?

Manfred slung the Gewehr over his shoulder. He glanced left to right – the trench was empty – and jumped down. He landed on the rotted wood of a trench board.

Manfred slid along the dirt wall. He slowed his pace two meters from the bend. Would the Mothmen be just around the corner? How many? One? Two? Ten?

He pushed down his worry. The mission was all that mattered. He took a soft, deliberate step, then another. Just one meter from the bend.

Something rustled behind him.

Manfred swung around, rifle raised.

THIRTY-TWO

The barrel of another rifle dominated Manfred's vision. His finger tensed around the trigger of his Gewehr. His gaze expanded until he took in the man behind the rifle.

"Herr Rittmeister." Heldmann, lying on his stomach at the edge of the trench, let out a short sigh and lowered his weapon. Two more men flanked him, Erhardt and Gude.

"We feared the worst when we saw your plane go down," said Gude.

"Fortunately, I came out of it in one piece. The same cannot be said for the Mothman who attacked me."

Heldmann grinned. He slid over the side and landed next to Manfred. Erhardt and Gude joined them.

Face scrunched, Heldmann looked up and down Manfred. "All due respect, *Herr Rittmeister,* but you are filthy."

"It is called camouflage." Eyebrow lifted, Manfred ran an appraising eye over the other pilot. "And your uniform certainly does not look freshly laundered."

Heldmann shrugged. "We did have to crawl for quite a ways, just to be safe."

"Mm-hmm. So did you land to rescue your leader?" asked Manfred.

"Actually, Erhardt spotted the monsters flying up from this trench." Heldmann nodded toward the other pilot. "We decided to land and try to pinpoint their positions for another artillery barrage." He paused, tilting his head to the left. "And also to see if our leader was still alive."

Manfred responded with a small smile. "Well, you've succeeded in the first part of your mission. Now let's complete the other part."

Erhardt, Heldmann, and Gude slung their rifles over their shoulders and drew their pistols. Those weapons were better suited for fighting in the confined space of the trench. But with his Luger lying somewhere in the fields of Flanders, Manfred had to make do with his rifle.

Hugging the dirt wall, he led the pilots through the zigzagging trenchline. He halted at every curve, sticking his head around it as far as he dared. No Mothmen were in sight.

The rotted wood trench boards sagged under his boots. Mud oozed between the cracks. In a few spots, it appeared as though a giant hand scooped out a portion of the trench. That had to be from artillery shells. In those parts the boards were gone. His boots sank into the damp ground below. Manfred stifled his groans as he pulled his feet free. He wanted to keep noise to an absolute minimum to avoid alerting the Mothmen to their presence.

The four rounded another curve. They found no winged monsters. The same with the next curve. Another large rut lay ahead. He led the three pilots through it. Like the other gouges, no wooden boards remained.

Again, the mud grabbed at Manfred's boots. He clutched the Gewehr in his left hand and pressed his right against the dirt wall to keep his balance. Jaw tight, he slowly pulled one foot out of the muck, then the other, and repeated the process with his next steps. He eyed the next curve in the trenches when his brain whispered a warning.

Up. Look up.

Manfred sneered at having to remind himself of that again. Unlike British or French infantrymen, the Mothmen could attack from the air as well as on the ground.

He raised his head. The sky was clear.

Manfred took another step forward.

Someone grunted. A dull thud of flesh on mud followed. Manfred swung around.

Erhardt lay on his side. "Dammit," he hissed, trying to push himself up with one hand.

"Careful how you walk," Manfred said in a harsh whisper.

"Yes, sir," Erhardt muttered.

Gude helped him to his feet, and the four slogged on. They went a few more meters before another line of trench boards appeared. Another curve lay just ahead. Manfred slid up to it and peeked around.

A few meters away, the left side of the trench wall had collapsed. Probably from a combination of a shell burst and the weather. It formed a dirt slope to the bottom. But it was the structure just beyond that caught most of his attention.

A squarish frame topped by a roof made of thick logs jutted out from the dirt wall. A bunker. Perhaps used for a headquarters or to store supplies.

Manfred slid back around the corner and looked at the other pilots. "There is a bunker on the other side. That would make a perfect hiding place for the Mothmen."

Heldmann's brow furrowed. "It would also be a tight fit, given how many of them came out of these trenches."

"True. Be that as it may, we have to check it out. Erhardt, come with me, and try not to slip and fall."

Erhardt winced. *"Ja, Herr Rittmeister."*

"Heldmann, Gude. Stay here and cover us."

The two men acknowledged the order.

Manfred pressed his back to the trench wall and slid along it. Erhardt did likewise. He focused on the bunker, then reminded himself to look up. No Mothmen flew overhead. None also emerged from the structure. He continued sliding along the wall.

Soon they reached the gouge in the trench. He tensed as his boots sank into the mud. Slowly, he pulled them out, creating a soft sucking sound. Manfred's throat seized, worried the Mothmen could hear it.

No winged monsters leapt out of the bunker.

He glanced back at Erhardt, who lifted his feet from the mud with care, staying upright.

Another step by Manfred, another slow lift of the foot from the damp earth. He repeated the process a few more times until he reached the next set of trench boards. His shoulders relaxed as he set a boot on the rotted wood.

A throaty warble carried through the trench.

He shot up his right hand. Erhardt froze.

The warbles continued, with some clicks and squawks mixed in. Manfred knew of no animal that could make such noises.

It had to be Mothmen.

He turned to Erhardt and tapped a finger against his ear. "Do you hear that?" he mouthed, then pointed at the bunker.

Erhardt swallowed, then nodded.

Manfred waved for him to follow. His heart hammered the closer he got to the bunker. Any moment he expected one of the beasts to appear.

But they remained in the bunker. Manfred thought he could make out three sets of . . . voices? Could this be their language? What could they be saying? Were they celebrating their kills in the sky? Plotting their next attack?

He reached the side of the frame. A tattered, filthy curtain hung over it. Manfred squatted, drew a deep breath to steel himself, and peeked around the edge.

His eyes bulged. Shock and disbelief clutched his mind.

Three Mothmen stood inside the bunker. Toward the rear of the space was a glowing blue circle, hovering just above the floor of wooden planks. It almost reached the ceiling and looked about two, two-and-a-half meters across.

The warbles and clicks of the Mothmen, along with the creatures themselves, faded from Manfred's attention. He could only gawk at the circle. His mind struggled to come up with a purpose for it. A source of light? Their version of a lamp? But it seemed too elaborate for that.

In the end, it proved beyond Manfred's comprehension. Simply another of the many mysteries surrounding these monsters.

If this belongs to the Mothmen, then it cannot be good for us . . . whatever it does.

Manfred started to withdraw when something shimmered within the blue circle. His eyes grew even wider as another Mothman stepped out of the light and into the bunker.

What? How? Manfred barely registered the inhuman noises coming from the new arrival's throat as he kept his gaze aimed at the circle. How could something just walk through it like that? It was just light. Not even light from say the glow of an electric bulb or a candle. Just . . . pure illumination. With no source to speak of. Yet this new Mothman emerged from it like a normal person walking through a door.

He nearly jerked when the thought hit him. Could it be . . .?

Manfred leaned away from the log that made up the doorframe and stood erect. Staring at Erhardt, he pointed to the bunker and held up four fingers. The other pilot drew a half-breath and nodded.

Manfred slung his rifle over his shoulder and took out a stick grenade. Erhardt did the same. Manfred slid along the dirt wall to the edge of the entrance. Taking a breath, he yanked the cord at the base of the stick, shoved back the torn curtain, and lobbed the grenade inside. He dashed past the bunker as Erhardt threw in his grenade and followed him.

Two muffled thumps echoed from within the bunker. Manfred tore the rifle from his shoulder as Erhardt brought up his pistol. The pair headed back to the structure . . .

And skidded to a halt when an agonized screech drilled into his ears. A shape fell against the curtain, tearing it free. A Mothman collapsed onto the trench floor, half-tangled in the soiled, ripped fabric. It wailed again, thrashing back and forth. Manfred saw blood staining its legs. Shrapnel from the grenades had wounded it.

Face stiff with determination, Erhardt stepped up to the monster. It cried out, trying to tear the curtain from its body.

Erhardt fired four times. The Mothman stopped moving. The pilot's gaze remained on the dead creature. Perhaps he felt great satisfaction in slaying one of these "demons." Did he see it as revenge for what happened to his sister years ago?

Whatever the case, the man did not have time to revel in his personal victory.

Manfred slapped Erhardt on the shoulder. The pilot spun around.

"Inside." Manfred went through the entrance.

Wisps of smoke floated above the floor where the other three Mothmen lay. Two did not move, blood spilling from several holes in their torsos and legs. The third one let out a raspy gargle. It struggled to push itself to all fours, bloody gashes on its left thigh and side.

Manfred lowered his rifle and put a round into the creature's head.

"Wh . . . What is that?" Erhardt spoke in a stunned whisper. His mouth hung open as he stared at the blue circle.

"It might be some kind of door," answered Manfred.

"Door?" Erhardt did not take his eyes off the circle.

"I saw a Mothman walk through it." Manfred shook his head. "It was the strangest thing. One second there was nothing. The next, it just appeared."

Erhardt glanced at him, then back to the glowing circle. "A doorway to Hell," he muttered.

"Perhaps. Perhaps not." Manfred stepped closer to it. Electric tingles ran up and down his body. The hair on the back of his neck stood on end.

"Are . . . are we going inside there?" The tremor in Erhardt's voice was unmistakable.

Understandable, thought Manfred. For someone as devout to their faith as his wingman, the thought of walking straight into Hell had to be terrifying.

But he need not have worried. "We are absolutely *not* going in there. For all we know, there could be hundreds of Mothmen on the other side. Thousands."

"Thousands?" The color drained from Erhardt's face. "What if they all come through? Our army would be pinned between them and the British. We could lose the entire salient."

Manfred turned around, pride surging within him. Despite Erhardt's fear, the man was still able to think, to consider the overall strategic picture. With some more experience and maturity, he could develop into a good leader.

"You are right, *Leutnant.* That cannot be allowed to happen."

"So what do we do?"

Manfred did not answer him, just gazed at the blue circle. His first instinct was to send some scouts in there to determine what exactly lay on the other side of that door. Probably more than some scouts. Maybe an entire infantry battalion. They'd also have to bring in scientists to study the glowing door.

And how long will all that take? The scientists might need a week or two to arrive here. What if the Mothmen launched a major attack before then? Also, would this discovery be enough to convince General von Armin to spare an entire battalion for a reconnaissance mission to who knows where with the situation along the Ypres Salient very much in doubt?

He stared at the mysterious door for a few more seconds, then stalked out of the bunker, Erhardt following. They met up with Heldmann and Gude, still guarding their rear against any Mothman attack. Manfred told them what he had found. The pair stared at him with mouths agape. Their wide eyes showed they had to fight to accept what their commander said.

"So what do we do with this . . . door?" asked Heldmann.

"We must close it as soon as possible." The words flew from Erhardt's mouth.

"And how do we do that?" blurted Gude. "No one has ever seen anything like this before."

"Erhardt is right," said Manfred. "That door must be closed."

"I'll ask again, sir. How?"

Manfred turned to Gude. "I do not know how we can close the door itself. But we can close the bunker it's in."

He sat on a fire-step, used by soldiers to shoot over a sandbag parapet, and took out a map from his case. He had Erhardt unfurl and hold it. Manfred then removed a small notebook and scribbled in it.

Putting his signature at the bottom, he stood and tore out the page. "Heldmann." He handed it to the pilot. "Take this and go back to your plane. Fly to the bridge where that mobile anti-aircraft gun is stationed. Have the unit there send a runner to our artillery battery. I want them to bombard these coordinates."

Manfred tapped the paper with his index finger. "A five-minute barrage should be enough to collapse the bunker. The Mothmen will never be able to get through the door when it's buried under dirt and wood."

"Ja, Herr Rittmeister."

"And if anyone questions you on this, tell them you are acting under my direct orders. And if anyone still questions you, tell them *I* am acting under the direct orders of General von Armin."

"Ja, Herr Rittmeister."

Heldmann put the paper in one of the inside pockets of his flight jacket. He then climbed onto a fire-step, clutched the edge, and tried to pull himself out of the trench.

"Here." Erhardt stood behind him, hands cupped.

"Danke." Heldmann put his right heel on Erhardt's hands. He pushed up.

Manfred watched the pair when something moved in the corner of his eye. He swung his head around.

Four Mothmen pounded out of the bunker.

THIRTY-THREE

Two Mothmen swung toward Manfred as he brought up his rifle and fired. Gude's pistol cracked. The monster at the forefront jerked and sagged against the dirt wall. It then slid to the floor of the trench. Its partner behind it staggered away, clutching its chest. It half-dove, half-fell into the bunker.

The other two Mothmen jumped into the air, flying out of the trench. Manfred fired his last round at them before pulling a five-round stripper clip from his flight jacket. Gude also reloaded his Luger.

Four more Mothmen rushed out of the bunker. One turned and charged toward them.

Erhardt and Heldmann hammered away with their pistols. Several red circles sprouted across the Mothman's torso and neck. Its legs wobbled before collapsing to its knees. Erhardt fired his last round. Crimson flew from the back of the creature's skull.

The other three Mothmen took to the air.

Another quartet of beasts followed. Manfred and Gude fired as Erhardt and Heldmann reloaded. One Mothman jerked, but still managed to take off. A second stumbled backward before retreating inside the bunker. The last two flew into the air unscathed.

Manfred yanked a grenade from his belt and flung it toward the bunker door. That should surprise the next squad that ran out. He and the other pilots ducked behind the curve in the trenchline.

The grenade thumped. Manfred looked around the corner and scowled. Wisps of smoke rose from the ground near the bunker. No Mothmen had charged out.

Until a few seconds later. Six bounded into the open and leapt into the sky. A cacophony of sharp pops surrounded Manfred as he and his pilots fired their weapons. Two of the monsters flinched, but their wings kept up a steady beat as they ascended.

Then he spotted the other eight Mothmen, more than two hundred meters above the trenches. Wheeling round and round as the newly arrived monsters joined them.

"What are they doing?" Erhardt craned his neck to gaze at the creatures.

"My guess is toying with us," replied Manfred. "Wanting to make us more afraid before they attack."

"I'll be damned if I give them the satisfaction." Heldmann sneered, determination burning in his eyes.

Manfred nodded. "That's how we all must think. We cannot allow them to make us so scared we freeze like back at our base." He fixed a hard gaze on his pilots. "If you start to feel afraid, ignore it and think only of killing these foul beasts."

Erhardt, Heldmann, and Gude all nodded. They unslung their rifles from their shoulders and fixed bayonets. Manfred ordered Erhardt and Heldman to take up position by the curve and cover the bunker. He stood on a nearby fire-step facing south. Gude got on another step two meters from Manfred's left and faced north. With a flying enemy like the Mothmen, they could attack from any direction. And certainly more would come through that blue circle in the bunker.

But how many? Manfred glanced at the bunker, then back up at the monsters in the air. They continued flying in a circle.

His heart banged against his chest. What were they waiting for? Maybe for more of their kind to surge through the strange door. Then they would dive on them. Surrounding them. No hope to escape. No hope of survival.

Stop it! A growl rumbled in Manfred's throat as he gripped his rifle tighter. That intense fear had been the Mothmen's doing, not his.

His gaze switched between the barrel of his Gewehr and the monsters overhead. He considered taking shots at them, but no. As he did in his plane, he wanted the enemy as close as possible to guarantee a hit.

The veins in his neck protruded. He felt the Mothmen would eventually get much closer than he wanted. Fourteen of the flying monsters against his rifle and Gude's. There would be no way they could shoot down all of them before they dropped into the trench. Manfred also doubted Erhardt and Heldmann could keep whatever creatures came out of the bunker at bay forever. The fighting would become hand-to-hand. Or more accurately, hand-to-claw.

Manfred quickly ran his gaze over the bayonet, then looked down at the cavalry sword hanging from his right side. The irony. He had distinguished himself in one of the most modern weapons in this great war. Now he may wind up fighting with one of the oldest weapons in history.

A piercing cry ripped through the air. Not one. Several. The wail went on, all of them merging into a chilling, hellish choir. Cold pricks crept up Manfred's spine.

"They are doing this to scare us," he said. "Do not let that happen. We are soldiers of the Imperial German Army. We will not shrink before the enemy, no matter what they are."

"*Ja, Herr Rittmeister!*" hollered Erhardt, his voice strong.

Heldmann and Gude responded in similar fashion.

The inhuman screaming continued above them, never letting up. It engulfed the very air, sank into Manfred's skull, into his very soul. How could any living thing make such a sound? How could they keep making it for so long?

Manfred shivered. Would it ever end? He wanted this damn noise to stop. Now!

He shut his eyes. *Do not give in. Fight it. It is just a noise. It is —*

"They're diving!" shouted Gude.

Manfred's eyes snapped open. All fourteen Mothmen dove on the trench, coming from all points of the compass.

He glanced at Erhardt and Heldmann. Both looked up at the approaching monsters.

"Cover the bunker!" Manfred ordered. "They will come from there, too."

The pair switched their attention back to the bunker, rifles ready.

Manfred grimaced. He could have used their rifles against the attacking Mothmen. But that could be what they wanted. For them to focus all their attention on this assault while an entire platoon of their comrades ran out of that blue circle.

He raised his rifle, picking out one creature. Manfred placed the iron sights on it as it barreled toward the ground. He wanted it closer . . . closer . . .

The rifle cracked.

The Mothman jerked to the left. It went limp and plummeted to earth.

Manfred pulled back the bolt and swung the rifle right. The other Mothmen pulled up, rolled, swerved, dove again, then pulled up and banked.

He targeted one of the monsters and fired. Miss. Another shot. Miss.

A second Mothman tumbled out of the sky. A hit by Gude. That *only* left a dozen.

Manfred fired again. A third miss. He tracked another monster. It swerved left, dove, then rolled right. He followed it in his sights trying to anticipate where it would turn next.

He pulled the trigger. The Mothman jerked. Its wings flapped like mad as it reached for its left leg.

Chamber empty, Manfred dug into his ammunition pouch for a stripper clip. He shoved it into the Gewehr.

Three Mothmen landed. Manfred aimed at the one in the middle and fired. Blood spurted out its chest as it toppled over. The other two creatures sprang into the air. Manfred aimed at the one to his right. The beast twisted to the side the moment he fired.

He grunted at the errant shot, then fired again. The Mothman grabbed its gut. Its wings flapped erratically as it tried to land. Manfred shot it again. The monster dropped to the ground.

Gunfire snapped to his right. He glanced at Erhardt and Heldmann. They let loose with shot after shot as several Mothmen rushed out of the bunker.

He clenched his teeth, wishing he could help them. But he had to stop the Mothmen in front of him.

Manfred fired and missed. The monsters jumped and flapped and twisted. The rifle barked again. The left wing of one Mothman twitched.

He reloaded the Gewehr as two thumps echoed through the trenches. Grenades from Erhardt and Heldmann. Bodies had piled up in front of the bunker. Mothmen pushed the corpses out of the way. Some were shot. Others managed to fly out of the trenches.

Manfred turned back to the field and burned through the five-round clip. Three more creatures fell. He pulled out another stripper clip as Gude fired his rifle. A Mothman at the edge of the trench across from them stiffened and tumbled over. It hit the rotted floorboards with a dull thump.

"I bet they are the ones scared now, eh, *Herr Rittmeister?"* Gude reached into his ammunition pouch.

Manfred nodded. "Then let's keep them scared."

Gude grinned and pushed the rounds into the chamber.

Dark, hairy hands grabbed the back of the pilot's jacket and yanked him out of the trench. He cried out as the rifle fell from his grasp.

"Gude!" Manfred hollered.

Gude lay on the ground, a Mothman towering over him. He scrambled back, reaching for his holster. The monster stomped toward him.

Manfred shot it twice in the back. It fell onto Gude's legs.

"Hang on." Manfred grabbed the lip of the trench and started to pull himself up.

Something moved to his left. He turned.

A Mothman dropped into the trench. Arms outstretched, it rushed him.

With a roar, Manfred thrust the rifle at the creature. The bayonet tore through its throat. Blood flowed down its torso as it went stiff.

Manfred pulled out the blade, then rammed it back into the Mothman's throat. The blood poured out in a red river. The eyes went as black as its hairy body. It collapsed.

More grenades burst nearby. Manfred ignored it as he swung back to Gude.

The breath stuck in his throat when he saw another Mothman over the pilot. Its hand whipped down over Gude. A fountain of blood rose and fell from his severed throat.

Anger and sorrow crashed into Manfred's soul.

The beast spun around and aimed its glowing eyes at him.

No time to mourn his pup. Manfred shot the monster twice in the chest. It crumpled next to Gude's body.

He gazed across the field, counting ten Mothmen on the ground. Half-a-dozen more soared overhead. His stomach dropped. The sweat on his body turned ice cold. One against sixteen.

His shoulders tensed when the realization hit him. Manfred checked over his shoulder.

Eight more Mothmen stood beyond the other side of the trench.

He glanced at Erhardt and Heldmann. They still fired on the Mothmen, who hurdled the bodies of their brethren on the ground in front of the bunker.

Manfred looked left, then right. The monsters stood their ground, eyeing the trench, as if knowing the end was near for him. Maybe they wanted to wait until he was terrified out of his mind, then they could all feast on his emotions before nearly ripping off his head as they had done to poor Gude.

But he was determined to not feel one ounce of fear, to deny them their feeding time.

He exhaled. He filled his mind with thoughts of Gude, Dr. Kretschmer, Ewers, Förster, Private Schaffner. All those killed or driven mad by these . . . demons. His rage burned, hotter and hotter by the second.

Manfred brought up his rifle and fired. The round burrowed into the chest of a Mothman, who dropped on its back.

He pulled out another stripper clip. It was all he could do now. Just keep firing until they –

The world whipped past. Something yanked his body into the air. A hard punch exploded across his back. He gritted his teeth and blinked, staring at the sky above. Three Mothmen flew above him. Grunting, he turned his head.

A Mothman stood less than two meters away.

Manfred rolled and got to one knee. He snapped his head left to right, looking for his rifle. It was nowhere in sight.

The creature did not move. Its intense red eyes bore into his. Probably trying to heighten his fear.

Eyes narrowed, Manfred stared back at it. The cracks of rifle fire and the thumps of grenades faded into the background. He was violating one of the tenets of *Dicta Boelcke.* Target fixation.

But with the odds so against him, what did it matter? Right now, this one Mothman made up his entire world.

The creature kept staring at him. Manfred did not look away. He had no idea how much time passed. It did not matter. He would stare at this demon for all eternity before he gave it the one thing it craved.

The creature trembled for a second. A noise rumbled from its throat, a mixture of boiling water and wooden spoons banging together. Could that be anger? Anger that he would not give into the fear it tried to provoke in him?

"You are not going to feed off me, you bastard."

The red eyes blazed more intensely.

"Not all of us are afraid of you." He slid his hand across his stomach, toward his sword.

The Mothman's shoulders rose. The boiling/banging noise it emitted grew louder.

Manfred pulled out his sword and sprang to both feet. The Mothman threw up its hands and shrieked. Manfred smiled. This time, *he* was making this monster afraid.

He rammed the blade into its chest. The Mothman arched its back. The red faded from its eyes. Manfred withdrew the blood-stained sword and let the beast crumple to the ground.

He looked around the field. A dozen Mothmen stared at him.

I always thought I would meet my fate in the sky. He could not win this fight. He might be able to strike down one or two before they overwhelmed him.

Manfred scowled briefly at the sword. He stabbed it into the ground and stepped back. None of the Mothmen moved. They just aimed their inhuman gazes at him.

"Come on. Let's finish this." He doubted they could understand him. Could they even detect the defiance in his voice?

It did not matter. They could see he was unarmed, that he was easy prey.

Or so they thought.

Almost as one, the monsters moved toward him.

Manfred moved his left hand to one of the grenades on his belt. His thumb hooked around the cord. When they got close enough, he'd pull it, and take as many of these beasts with him as possible.

Step by step, the Mothmen neared him. Manfred stood erect. He accepted his fate, the fate of a true soldier. No fear. Nothing for them to feed on.

The Mothmen continued their steady advance.

Manfred's thumb tensed. Any second . . .

Gunfire rattled around him. He slipped his thumb out of the cord and dove for the ground. Three monsters jerked and fell.

A thunderclap rocked the air around him and shook the ground. A gusher of dirt shot in the air behind the Mothmen. Some of the creatures took to the sky. Others threw themselves flat to the ground. Where was this fire coming from?

Something boomed to his left, followed by several sharp pops. Manfred turned his head.

The anti-aircraft truck that had been guarding the river sat nearby. Six soldiers charged toward him, led by a familiar figure. Werner Voss!

Manfred could not help but smile. He had seen his friend flying east as he made his way to the trench. He must have contacted the anti-aircraft unit and got them to come here.

Voss and the soldiers fired from knees or standing positions. Two Mothmen fell from the sky. The remaining ones on the ground scrambled to their feet and took to the air. One threw out its arms and dropped to the ground, a bloody hole in its back.

The 7.7cm gun in the bed of the truck boomed again, shooting at the flying creatures. The shell missed. Four Mothmen dove at the vehicle.

Another soldier took a knee close to the truck, clutching a long-barreled Bergmann light machine gun. He raised it at the attacking creatures and pulled the trigger. The steady chatter carried across the field. One by one, the Mothmen spiraled from the sky.

"*Herr Rittmeister!*" Voss hurried over to him.

Manfred pushed himself to his feet by the time the commander of *Jasta* 10 reached him.

"*Herr Rittmeister.* Are you all right?"

"I am now." Manfred clasped Voss on the shoulder. "You got here just in time."

Voss beamed. "I saw some of those ugly bastards flying out of the trench. When you went down, and when Erhardt and the others went after you, I figured you would need reinforcements in case you ran into more Mothmen."

Manfred glanced at the truck, the gun crew firing again at the airborne creatures. "I'm glad you were able to convince them to leave their position to come here."

Voss shrugged. "It was not difficult. I told them if they let *Der Rote Kampfflieger* die, they would personally have to answer to the Kaiser himself. That got their asses moving."

Manfred grinned, grateful for his friend's force of personality. "Come. Erhardt and Heldmann are still in the trenches fighting the Mothmen." He yanked the sword from the ground and took off running.

"What about Gude?"

"A Mothman got him." Manfred grimaced, not breaking stride.

No reply from Voss. Probably digesting the news of Gude's death.

The pair reached the bend in the trenches leading to the bunker. Manfred gritted his teeth as he neared the edge, dreading the worst.

He let out a long breath of relief as Erhardt and Heldmann stared up at him. Blood coated both their bayonets. Manfred counted half-a-dozen Mothmen bodies near the two pilots. Other creatures lay dead along the trench floor leading back to the bunker.

"Voss?" Erhardt eyed him. "Where did you come from?"

"I was getting help for you." Voss eyed the dead monsters in the trench. "Not that it looks like you needed it."

"Believe me, we did," said Heldmann. "We were getting low on ammunition."

"Well let's get --"

Manfred stopped when he saw Voss raise his rifle. Three Mothmen soared over the trench less than eight meters from them. But the beasts ignored him and his pilots. They pulled up, dropped into the trench, and darted into the bunker. Two more Mothmen descended toward the trenches.

Two soldiers rushed past Manfred and Voss. One carried a long nozzle, the other had a cylindrical tank on his back.

A flamethrower team. The weapon had been around for a couple of years. Manfred had never seen it in action, but heard stories about it. Horrifying stories of what it could do.

The two Mothmen pulled up, flapping their wings, and preparing to land. An evil hiss flowed from the flamethrower. A stream of fire followed. Both creatures turned just as the mass of orange and yellow enveloped them. They dropped to the ground, flailing. Both unleashed high-pitched wails that sent a wave of cold pricks up Manfred's spine.

One of the burning Mothmen tumbled over the edge of the trench. The other lay still on the ground, flames rising from its body. Manfred grimaced as the stench of burnt hair and flesh crept into his nose.

He checked the sky. It was clear of Mothmen. As for the ground, all he could see were corpses of the monsters.

"It looks like we've won," Voss said with a huge smile.

"Not yet." Manfred shook his head and walked over to the anti-aircraft truck.

"Who is in charge?" he asked the gun crew.

"I am, sir." A short soldier came to attention. *"Unteroffizier* Gitner."

"You did a good job using this gun like a standard artillery piece."

"Danke, Herr Rittmeister."

"Now I need you to do it again." Manfred pointed toward the trenches. "I need you to shell that bunker until it collapses. There is a slope you can park your truck on that will give you an almost point blank shot at it. We bring down that bunker, we trap any Mothmen inside."

Gitner checked around Manfred's shoulder and furrowed his brow. "All the Mothmen are coming from that bunker? But there were so many."

"I know it sounds unbelievable, *Unteroffizier,* but it is true. Now position your truck on that slope and fire on that bunker until it collapses. Otherwise we could have to fight a hundred of those damn beasts."

Gitner drew back his head, eyebrows scrunched. "A hundred? But how --"

"I have no time to waste with questions," snapped Manfred. "Carry out my orders."

"Er . . . yes, sir."

The *unteroffizier* assembled his crew and directed the driver toward the slope.

"All those Mothmen came from that one bunker?" Voss tilted his head to one side in a doubtful expression.

Manfred nodded. "They came through some sort of glowing door inside there."

Voss said nothing, just raised an eyebrow.

Manfred sighed. "I know. It sounds unbelievable."

"It does. But everything about these last few weeks has been unbelievable, so what's one more bit of strangeness?"

Grunting in acknowledgment, Manfred turned to the truck. It went down the slope only about a meter before stopping. The crew swung the gun toward the bunker. Manfred gripped the hilt of his sword tight, half-expecting more Mothmen to appear.

The big gun thumped. Wood and dirt flew into the air. The crew slammed home another shell and fired again . . . and again . . . and again. After half-a-dozen explosions, the mangled roof of logs collapsed. Dirt cascaded into the ruined bunker like a thick, earthen waterfall.

Manfred and Voss headed over, joined by Erhardt and Heldmann. He stared at the mound of dirt and shattered logs that had buried the blue circle.

"Ha!" barked Voss. "I'd like to see the bastards dig themselves out of that mess."

Manfred nodded slightly and continued to stare at the mound. Relief flowed through him. The Mothmen would not be able to use this doorway to attack them.

We've beaten them.

Even as he thought that, a whisper came from the back of his mind. They had blocked this doorway. But if there was one, could there not be others?

THIRTY-FOUR

"So this . . . doorway of theirs is closed. There is no risk of them attacking us again."

Manfred noted the hopeful tone in General von Armin's voice, wishing he could share that feeling. "I don't know if I would say that, sir. There is always the possibility the Mothmen could dig themselves out of that debris."

The commander of 4th Army grunted, resting his clasped hands on his desk. His eyebrows scrunched together for a moment before staring back at Manfred, seated in front of the desk in von Armin's office.

"I will send a company of pioneers to those trenches and tell them to seal the remains of that bunker with whatever they have. Bricks, concrete." The general gave a brief, half-smile. "I doubt those monsters will be able to dig through that."

Manfred said nothing, just nodded. *I hope so.*

"Then this matter is settled." Von Armin's chin rose in a confident manner. "The Mothmen have been defeated. Now we can focus all our efforts on stopping the British offensive in the Ypres Salient."

The general squared his shoulders. "*Rittmeister* von Richthofen. You and your pilots performed with exceptional skill and courage in the face of perhaps the most extraordinary threat our empire has ever faced. Your efforts may have averted a disaster for our forces along the salient, and for that, I commend you and your men."

"Thank you, sir." Manfred got to his feet when von Armin did and shook his hand when he offered it.

"Now you and *Jagdgeschwader* One can forget about monsters and devote all your attention to shooting down British and French planes."

"Yes, sir. But if I may, there is something I'd like to show you."

"What is it?" Von Armin lifted an eyebrow in apparent curiosity.

Manfred opened a satchel and took out a sheaf of papers. "When I returned to base, I got together with my *jasta* leaders. We went over our battles with the Mothmen and came up with new doctrine on how to fight them."

"You did?" von Armin's voice was low, flat.

"I do not believe this is the last time we will see these creatures. Yes, we may seal that doorway near Meulebeke, but it is possible there are other doorways. Not only in Belgium, but other places around the world. For all we know, there could be one in Germany."

Von Armin grimaced.

"We need to be prepared the next time they appear. We need to have a strategy." Manfred held up the papers.

The general stared at him in silence for a few seconds. Manfred's brow wrinkled. What was taking him so long to respond?

Finally, he said, "May I see?" and reached out his hand.

Manfred handed him the papers.

Von Armin looked over one sheet, then another, then a third. He gave a slight grunt and took a few steps away from his desk. He stood next to a metal waste bin . . .

And dumped the papers into it.

"Sir?" Manfred's mouth hung open in shock. Why would the general do that?

Von Armin said nothing. He grabbed a book of matches off the desk and lit one.

Manfred's eyes widened as the match fell into the bin. He watched in stunned silence as flames crept up the papers, the doctrine he, Voss, von Döring, and Dostler spent hours discussing, writing, and refining.

All of it burning before him.

"Sir, why?" He pointed a hand at the bin. "That will be valuable the next time we fight the Mothmen."

"There will not be a next time, *Rittmeister.*" Von Armin's words came out sharper than a sword.

"But, sir --"

The general cut him off. "Listen to me, *Rittmeister*. The Mothmen *do not* exist. Your battles with them never happened."

Manfred could only blink as he tried, and failed, to comprehend what he had just heard. It took a few moments to fight through his disbelief and find his voice. "I . . . I do not understand."

Von Armin's scowl deepened. "Then I shall enlighten you."

***

Manfred bellowed and banged a fist against the side of his cockpit as he flew over Flanders. How he contained his roiling fury while von Armin spoke, and during his entire walk through HQ and to his plane, he had no idea. But as soon as his Albatros DIII – borrowed from Erhardt until he could get a replacement plane for his wrecked D.V – it exploded from his mouth with all the power of Krakatoa. Primal yells, and words he rarely used. Foul words the lowliest of foot soldiers would spout without a second thought. Not a Prussian officer of his standing.

But the discipline he had prided himself on failed. No, it did not fail. He discarded it. Curses continued to fly past his lips. Many of them aimed at General von Armin. Even in private he would never conceive of saying such things about a superior officer. But after everything he'd been through the past few weeks . . .

"Blind, addled, shit-brained ass!"

His yelling abated the closer he got to base. By the time he landed, Manfred had regained control of his emotions. He was still furious, but at least he could manage it better.

A good thing, too. He could not come off like a screaming madman before his men when he addressed them. *Much as I want to.*

He sat in the cockpit until his splitting headache and swirling nausea subsided, then dispatched a couple of the mechanics to tell the pilots to gather in the mess tent in thirty minutes. Manfred went back to his tent, where Moritz greeted him with a wagging of the tail and a lick of the hand. He rubbed the great dane's head. Not only did the dog appreciate it – judging by the way its tail wagged harder – but it also had a calming effect on Manfred. It wasn't long before he completely felt like his old self, ready to speak before his men in a manner befitting his position.

Five minutes before his briefing was to begin, he headed to the mess tent. All the pilots of *Jagdgeschwader 1* were there, and came to attention when he entered.

Manfred waved them to be seated and stood in front of them.

"I have just returned from meeting with General von Armin. He has personally commended us for defeating the Mothmen and preventing them from disrupting our operations in the Ypres Salient."

Several of the pilots nodded or smiled. Receiving praise like that from a general was rare, and should be celebrated.

Manfred also wanted to start off with something positive before delivering news sure to infuriate them.

"I also discussed with him the concerns we all share, that the Mothmen could return. I presented him the doctrine that I and *jasta* leaders Voss, von Döring, and Dostler crafted."

He took a deliberate breath before continuing. "The general threw it in a waste bin and set it on fire."

The pilots turned to one another, many with shocked expressions, some muttering in astonishment.

"Why would he do that?" blurted Dostler.

"Because according to the general, the Mothmen do not exist."

That brought more shocked looks and mutters.

"How can he just dismiss them like that?" Erhardt flung an angry hand out in front of him.

"If the Mothmen don't exist, then what the hell were we fighting the past few weeks?" said Voss.

"This makes no sense," Heldmann chimed in.

"As General von Armin told me," Manfred began, "given the unusual nature of the Mothmen, if their existence became common knowledge, it could terrify both our troops on the frontline and the civilians back home. It could demoralize them to the point we risk losing the war."

Jaw tight, he gazed at his pups before speaking. "Our superiors do not simply want to keep the Mothmen a secret, they want them completely erased. Forgotten, lest someone accidentally reveal the truth to the rest of the world."

"But *we* saw them." Erhardt came forward in his seat, hands extended as though pleading. "Fought them. We wrote about them in our reports. We even have a photograph of one."

"Any mention of the Mothmen will be stricken from all official reports, and the photograph is to be destroyed. Anyone in this unit who discusses the Mothmen with anyone will not only be subject to court martial, but will be committed to an asylum and deemed insane."

Manfred studied the faces of his pilots. Some appeared shocked, some concerned, and some angry.

"In addition," he continued, "our battle with the Mothmen near Meulebeke has been purged from official records. It never happened. Therefore, any pilot who shot down an enemy plane will not be credited with the kill."

The grumbling grew louder. Manfred wanted to join in. He had brought down two fighters over Meulebeke. Kills number fifty-eight and fifty-nine. Now denied to him.

"And what about Gude?" Voss demanded with an acidic tone. "Is he purged from the records? Him and the other pilots we lost?"

Manfred dipped his head, scowling. "Gude and the others who died in that battle will instead be listed as having died in non-combat plane crashes."

The men howled with anger and disgust.

"Disgraceful!" blared Hans Adam.

"They are dishonoring them!" hollered von Döring.

Manfred gave them a minute to get out their fury . . . and to push down his.

"I cannot believe the general would do this." Erhardt's body trembled with anger.

"It is not just General von Armin," said Manfred. "These orders come directly from Field Marshal von Hindenburg and the Kaiser himself."

That announcement did not bring about the same sort of shouts as the other ones. The men still seethed, but held their tongues. As upset as they might be, jeering the Chief of the German Greater General Staff and the ruler of the German Empire was a line they dared not cross.

"So that is it?" Erhardt's face reddened. "We are supposed to forget about those demons."

"Yes." Manfred nodded. "Those are our orders."

His shoulders tensed and he held his breath. What he was about to say was something he never would have conceived of doing before now. Part of him still felt it was wrong. It went against everything he'd been taught since entering military school at eleven years old.

But given the threat still posed by the Mothmen, he could not simply forget about them.

"Now here are my orders."

Every gaze in the tent fell on him. The men stayed silent as he went on.

"As much as they want, our superiors cannot simply wish these creatures away. We certainly cannot forget about them, and we must be ready for them."

Manfred stood as straight as possible. "We must keep our eyes and ears open, and be alert for any word about the Mothmen's return. Any report, any whisper of some . . . unidentified flying creature. You will bring it to the rest of us, and we will act on it."

"How can we do that without our superiors knowing about it?" asked Dostler.

"We will have to find a way. We will have to become . . . silent hunters. Operating in the shadows to protect the empire . . . perhaps the entire world from these monsters. Whether it is during this war, or after it, or in the next war."

"Are you sure about a next war?" Voss asked with a wry grin. "Didn't that English writer Wells call this 'The War That Will End War?'"

Manfred scoffed. "Do you truly believe that?" He stared around the tent. "Does anyone truly believe that?"

No one answered. Several shook their heads.

"War is part of human nature," said Manfred. "There will be other wars after this so-called 'War That Will End War.' For all we know, the Mothmen are drawn to conflict. What better place to feed on fear than a place where millions of men walk around with death hanging over their heads every minute?"

Manfred let out a long breath, stared at the ground for a moment, then back at his pups. "I am sure you are aware what I have just suggested is insubordination. Not only that, but insubordination in time of war. If General von Armin learns of this, I will be court martialed and probably face the most severe of punishments. That goes for any of you who choose to follow me in this. So I offer you a choice. If you are concerned about your future, your very well-being, you may leave this tent, and I will think no less of you. But if you believe the Mothmen continue to pose a threat to the German Empire, then stay here, and when they appear again, we shall fight them."

Erhardt practically jumped to his feet. Voss followed a split-second later.

Heldmann stood next, then Adam. Von Döring and Dostler stared at each other for a few moments, then got to their feet.

In less than a minute, the pilots of *Jagdgeschwader 1* stood as one. Not a single man had walked out of the tent.

A smile spread across Manfred's face. Pride filled his soul. Never did he think he would feel that way toward a group of men who defied an order from the Kaiser.

But if that was the only way they could protect Germany from the Mothmen, so be it.

THIRTY-FIVE

Point Pleasant, West Virginia
December 15th, 1967

"*This is your top of the hour news from WCMI Radio in Huntington. I'm Lester Hubbs.*

"*Rescue efforts continue at this hour in Point Pleasant at the site of the Silver Bridge, which collapsed earlier this evening. It is estimated that dozens of cars plunged into the freezing waters of the Ohio River. Authorities in charge of the rescue efforts say the number of dead and injured is not known at this time. Mayor D.B. Morgan says he has contacted Governor Smith of West Virginia and Governor Rhodes of Ohio to request more help to recover the people who fell in the river.*"

Retired *Luftwaffe* Lieutenant General Fritz Erhardt's shoulders sagged as he stared through the car's windshield, failure digging deeper into his insides.

Three times over the past year he'd come to this part of the United States. Ever since the first articles about the sightings appeared in the newspapers. They had even called the creature Mothman, just like he and his fellow pilots had fifty years ago. Not much of a coincidence, given their appearance.

But the American press treated it like a myth, something akin to a ghost story. A distraction from all the political and social unrest in this country.

Erhardt, however, knew the truth. The Mothmen were back, and he had to destroy them.

I should have done that months ago. But none of his previous investigations in West Virginia turned up any evidence of the creatures' nest. Perhaps if he'd stuck with it longer instead of returning to West Germany when the trail grew cold each time, he might have prevented what happened to the Silver Bridge.

Erhardt snorted and stared out the passenger window at the darkened trees along the rural road. He couldn't blame himself for one year of fruitless searching. He'd had fifty years to find the Mothmen's home.

But there had only been so much he could do to track down the beasts during the madness that had consumed the first half of this century. Once the First World War ended, Erhardt thought he could devote time to learning what the Mothmen were and where they came from. The Kaiser had abdicated the throne and fled to the Netherlands, the monarchy was no more, so Erhardt no longer felt obligated to follow the former emperor's orders when it came to staying silent about the creatures.

But the war had left Germany broken physically, spiritually, and economically. Survival took precedence over the search for monsters.

He grimaced, thinking back to those early post-war years, working in a factory and being paid in marks that would be worthless mere hours after he stuffed them in his pocket due to the out-of-control inflation. Struggling every day to buy food and clothes and trying – sometimes unsuccessfully – to keep his apartment.

Just as the country started to recover from the war, the Great Depression hit. Once again, the search for the Mothmen was pushed aside as keeping his wife and two sons fed became his priority. But Germany rebounded once Hitler came to power. With the air force reconstituted, experienced men were needed to lead it.

Erhardt thought back all those years ago, staring at himself in the mirror in his new uniform, pride flowing through him. He had quickly risen through the ranks, becoming a *generalmajor* by the late 1930s. Now that he had power and a large organization behind him, he could finally conduct a proper search for the Mothmen.

Or I should have had the courage to do so. Perhaps I could have prevented that bridge from collapsing.

He lowered his head as the car sped along the road. He feared the strict Nazi regime would never have believed his story of battling flying monsters over Flanders, especially with no proof.

Erhardt had searched *Luftwaffe* archives for any reports or photos of the Mothmen. He could not believe the General Staff would erase all records of its existence. They would keep at least one report, just in case the creatures returned.

But he found nothing. Either the generals had purged anything having to do with the Mothmen or those reports disappeared in the chaos of the government transitioning from the monarchy to a republic. He also had no idea what happened to the bodies recovered from Flanders. They had either vanished or were disposed of on General von Armin's orders. Whatever the case, he could not tell his superiors of the Mothmen unless he wanted to be booted out of the service.

Luck had been with him as he reconnected with his old squadronmate Alois Heldmann, serving as the inspector of a flight school. Together they found a handful of *Luftwaffe* personnel willing to believe them, or who themselves had witnessed strange creatures and phenomena. Behind the regime's back, they became *Die Schatten Jäger* – the Shadow Hunters. Honoring *Rittmeister* von Richthofen's orders, they became silent hunters, operating from the shadows, watching, waiting for the Mothmen to return.

Then World War II broke out.

Erhardt did what he could to find the monsters, but the majority of his time was spent trying to protect German-held oil fields in Romania from Allied bombers, then fleeing from the advancing Red Army.

Then mourning the death of his youngest son.

A lump formed in his throat, as it did every time he thought of Theodor. Sent into the air to face waves of American bombers and their escorts with the most rudimentary of training. He had lasted three missions before being shot down by a Mustang.

His cheek twitched. He gritted his teeth, holding back his emotions. He could not afford to show it in front of the other men in the car.

Erhardt eyed the road ahead, lit up by headlights, his mind drifting back more than twenty years. The end of the war. Imprisoned for months until the Allies determined he was not a war criminal. Then back to a Germany broken physically, spiritually, and economically, and also divided. Just like after the First World War, he did what he could to care for his family.

Ten years later, he found himself unexpectedly back in uniform. NATO decided West Germany needed its own armed forces to help fight the Soviets should they decide to invade. The air force was again reformed, and again they needed experienced men to lead it. Never having been a member of the Nazi Party, never having been linked to any atrocities during the war, Erhardt was allowed to join.

With West Germany stable, with this new "war" cold and not hot, he and Heldmann – now a civilian – rebuilt *Die Schatten Jäger*. They could finally dedicate the necessary time and resources to find the Mothmen, and any other mysterious creatures out there.

"There's the power plant." The burly driver of the station wagon nodded ahead of him.

Erhardt glanced at Master Sergeant Bob Kirkland, a U.S. Air Force Security Policeman, then stared down the road. The silhouette of a square-shaped building stood out in the darkness.

"Are you sure they are not in there, sir?" asked the slender man sitting in the backseat. "That would seem a good place to have their nest."

Erhardt looked over his shoulder at *Oberleutnant* Bernhard Weber. The F-104 pilot had become a *Schatten Jäger* two years ago after pursuing a UFO. "I explored the building during my previous investigations. There was no evidence the Mothmen were using it. Besides, it is my experience that they do not use rather obvious places to hide."

"One incident does not a pattern make," said the lean, older man sitting next to Weber.

"Do you have a better theory, my friend?" Erhardt replied to Brigadier General Logan Jones.

His American counterpart gave him a half-grin. "If I did, I would have told you already."

With a brief smile of his own, Erhardt faced forward and drew a long breath. Anticipation swelled within him. So did fear. He pushed it down as much as he could. He would not give these beasts the satisfaction of scaring him, or feeding off that emotion.

Kirkland pulled into the parking lot of the abandoned power plant. Erhardt scanned the windows and rooftop for any winged shadows. The Mothmen may not be using the place for a nest, but they could have posted a sentry for all he knew.

He saw no sign of the creatures.

The station wagon stopped and Kirkland killed the headlights. Erhardt opened the door and slipped out of the car. Despite his thick coat and woolen cap, he shivered as the cold night air enveloped him and dug into his bones.

He headed to the rear of the vehicle, as did Weber, Kirkland, Jones, and the final member of the group. Alois Heldmann. The American Air Force general swung open the rear door and threw back a blanket covering the cargo hold. Erhardt nodded with satisfaction.

Before him were three stubby M3 submachine guns and several rectangular, 30-round magazines. The cargo bay also held two M870 shotguns with ammunition bandoliers, five M1911 pistols with extra magazines, five satchel charges, and a wooden crate of hand grenades.

Erhardt turned to Jones. "I never imagined befriending an American bomber pilot would prove so useful."

Jones grunted. "So now the truth comes out. I feel so used."

"I'm sure you will get over it." Erhardt softly chuckled. His friendship with the general had been unexpected. Unimaginable would probably be a better word.

Jones had been captured after his B-24 was shot down over Bucharest in 1944. During his interrogation, he had mentioned seeing "strange lights" during the mission. When the report reached Erhardt's desk, he met with the American pilot thinking these "Foo Fighters" as the Allies called them might be linked to the Mothmen.

He found no evidence of that, but despite being on opposite sides of the war, Jones was appreciative that someone outside his crew believed him about this phenomenon.

"Even if you are a Kraut," he had added.

The two had kept in touch after the war, with Jones forming his own version of the Shadow Hunters in the U.S.

Heldmann and Jones took the shotguns, while Erhardt, Weber, and Kirkland grabbed the M3s. Each man helped himself to the pistols, the satchel charges, and a few grenades. Jones had requisitioned the weapons from a National Guard armory for "top secret" reasons. As a general and former head of Project Blue Book, he had the rank and necessary clearance to get whatever he wanted without the need to explain himself.

Erhardt ran his eyes over the submachine gun, clenching his teeth. He had planned on scouting the area while General Jones brought in reinforcements before making their move two days from now. Then the Silver Bridge collapsed. The Mothmen had to be responsible. Residents had been seeing them for a year leading up to today's disaster. This could not be a coincidence.

Who knew what other chaos they had planned to terrorize the people of Point Pleasant and feed off their fear? Their little group could not afford to wait for more men. They had to move now.

The five checked their weapons and headed into the woods. Kirkland took point, the M3 in one hand, a flashlight in the other. Heldmann also carried a flashlight. Erhardt grimaced for a second. The beams would definitely give them away. But they could not afford to stumble around the woods in absolute darkness and trip over something or fall into a freezing pond.

Of course, it was believed the Mothmen had natural night vision, so what did it matter if his group used flashlights or not? The creatures would still see them.

They snaked through the woods, weaving around stumps and ponds. They did their best not to step on any dry leaves or fallen branches, their crackling possibly alerting the Mothmen.

If they are even here.

No. They had to be here. Several people had witnessed the creatures in what was called the TNT Area of Point Pleasant.

Erhardt caught sight of a brush-covered concrete dome. It was one of many "igloos" that made up the TNT Area. The place had been used to manufacture and store munitions during World War II.

His eyes lingered on it, wondering how many of the bombs and shells made here had been used to pound Germany into rubble. Had some of the bombs Jones dropped from his B-24 come from this place?

He shook his head, ridding himself of the thoughts. The war had ended twenty-two years ago. His country was free – at least the half of it he lived in – and had been rebuilt. General Jones was his ally, not his enemy. Their mutual enemy hid somewhere in these woods.

"Remember to keep looking up," he told the others. "Mothmen can attack from the ground *and* the air." To emphasize the point, he lifted his head and peered at the night sky. It was clear of flying monsters . . . as far as he could tell.

The others did the same, alternating their gazes from ground level to above them.

They skirted more ponds and trees, and passed more igloos. Erhardt had examined them during his previous times in Point Pleasant. The steel doors remained locked and secured. There was no way the Mothmen could have opened them.

They had gone about a mile before needing a break. More like he and Heldmann needed a break. Erhardt maintained a daily regimen of walking and swimming to keep fit. But no matter how good of shape he was in, seventy-two was still seventy-two.

After a five-minute rest, the group resumed its trek. They put another mile under them before he and Heldmann needed another break. Pain drilled into his back and squeezed his ankles and knees. His daily walks did not include lugging around ammunition and explosives. He also usually walked on flat surfaces, not the uneven terrain of a forest.

"You starting to regret coming out here?" asked Jones, who sat against a tree across from him.

"Absolutely not. This is where I need to be." He grimaced and rubbed his left shoulder, trying to rid himself of the pulsating pain from carrying a loaded backpack.

"Your son disagreed," Heldmann chimed in.

"Wolfgang did not fight the Mothmen. I did."

"Wolfgang is in charge of the *Schatten Jäger* now," Jones pointed out.

"Well, I am still his father," said Erhardt. "So that carries more weight."

Jones shrugged. "He's just trying to look out for you, Fritz. Let's face it, you're no spring chicken."

"Neither are you."

"Maybe, but I'm still about twenty years younger."

Erhardt groaned and rubbed his shoulder again, cracked his back, then stretched his legs. After a ten-minute rest, they set out again.

Then took another break after less than a mile. Erhardt's shoulders and back ached to the point rubbing them did nothing. He lowered his head, thinking about the argument he'd had with Wolfgang before leaving for America. It had been similar to his previous three missions to West Virginia. His son reminded him he was old and retired. He also accused him of being obsessed with the Mothmen.

Perhaps he was. But Wolfgang had only grown up hearing stories from him about the monsters. He had not actually battled them. He never had a Mothman reach into his soul and grip him with fear just to feed off it.

Erhardt could not just sit back while these demons terrorized and killed more people. He also had to keep his promise to

Rittmeister von Richthofen, that he would always stay alert for the Mothmen's return and do what he could to stop them. There was nothing in "The Red Baron's" oath that freed him of that obligation once he reached a certain age.

It took much effort on his part – often loud effort – but he had convinced Wolfgang to let him undertake this mission. Guilt had clung to him during his flight to the U.S. and throughout his search for the monsters. If the worst happened, he did not want his last words to Wolfgang to have been spoken in anger.

I should have called him, made amends. But tracking down the Mothmen had taken up a lot of his time.

You could not spare just a few minutes for a phone call?

Erhardt sighed as he slowly pushed himself to his feet to resume their hike. He glanced over at *Oberleutnant* Weber, who looked none the worse for wear.

Maybe his son was right, thought Erhardt. War, and monster hunting, were endeavors better suited for the young.

He shook his head. Old or not, he had to be here.

They pushed on for another half-mile before coming upon a ravine. Erhardt took out his map and had Heldmann shine his light on it. "This should be it."

He looked over the hand-drawn map, taking a deep breath. The day before he'd passed himself off as a magazine reporter to interview a teenage couple who had seen a Mothman. The girl said all the sightings had scared the local youth to the point they stopped having cave parties. When he'd asked what those were, the boy explained a few times during the summer they and their friends would go to a cave in the woods a few miles from the TNT Area to, "have fun."

Meaning drink and screw.

The girl, a fairly good artist, had drawn a map to the cave. It was the perfect hiding place for the Mothmen. They had to be holed up there.

Erhardt slowly shook his head, thinking of the Silver Bridge. *If I'd only interviewed them a few days earlier.*

He grunted, pushing aside his guilt. He needed to focus on the task at hand.

Erhardt stepped over to Kirkland. "Master Sergeant. Shine your light there." He pointed down into the ravine.

"Yes, sir."

Kirkland ran his flashlight over the slope until it settled on a small fallen tree.

"That's some half-assed camouflage," said the big Air Force policeman.

Erhardt nodded and waved the men ahead. They stepped carefully down the slope, leaped over the small stream running through the ravine, and walked up to the fallen tree. Erhardt scanned the sky again. No Mothmen.

Kirkland and Weber pushed aside the tree. Erhardt, Heldmann, and Jones all had their weapons ready. With the *Luftwaffe* jet pilot covering him, Kirkland shone his light inside the cave.

"Clear," he reported.

Erhardt breathed halfway and held it. The discomfort, the bitter cold, it all vanished from his consciousness. All he could think about was slaying the Mothmen again.

With Master Sergeant Kirkland on point, the group entered the cave.

They hadn't gone far when they neared a bend to the right. Erhardt scanned the walls. With the white beam blazing ahead of him, it was impossible for him to determine if his hunch was correct.

"Master Sergeant," he spoke in a whisper. "Turn off your light."

Kirkland turned to him. "Sir?" His tone was unsure.

"Trust me."

The big man's lips pressed together for a moment. "Yes, sir." The beam cut off.

"Look," Weber said in an urgent whisper as he pointed.

A hazy blue glow shimmered on the cave wall.

Heldmann crept up to Erhardt. "It's them, that's for sure."

Erhardt nodded. "Move out."

Kirkland checked the corner. "Clear."

They moved around the bend. The glow became brighter as they approached another bend. All five stepped carefully, trying to make as little noise as possible. Erhardt did not know how many Mothmen could be here. He doubted there was just one as the American newspapers suggested. He half-expected to be outnumbered, like they had been in the trenches in Flanders in 1917. But they had survived that battle. Everyone except Gude. He also had much better weapons now than he did back then.

Kirkland reached the corner and peeked around. He looked back at them and used hand signals to indicate he saw four Mothmen.

Erhardt quietly stepped over to the master sergeant. The bigger man moved aside, allowing Erhardt to stare around the corner.

All his muscles tensed. His mind dragged him back to the First World War, but he put a halt to it. He needed to focus on the present day.

Four Mothmen sat against the wall of this alcove, two to the left, two to the right. Between them was the blue portal.

The creatures barely moved. Two of them emitted low noises, something between a groan and a click.

Erhardt grinned. He could guess what happened to them.

He backed away from the corner and gathered the others.

"It might be me, sir," Kirkland spoke in a whisper, "but they look out of it. Kinda like those hippies when they do too much drugs."

"An excellent analogy, Master Sergeant. Given how many people were on that bridge when it collapsed, the Mothmen likely overdosed on all that fear."

"Then we should have no problem taking out the ugly bastards," said Jones.

The plan of attack was simple. Kirkland and Weber would rush out to the left, while Erhardt, Heldmann, and Jones stayed near the bend. Then they'd fire until all those monsters were dead.

Kirkland took one last look at the Mothmen, then turned to the others and held up three fingers on his left hand.

Three . . . Two . . . One.

Kirkland and Weber rushed toward the other side of the cave. Erhardt, Heldman, and Jones jumped out into the open. Two of the Mothmen turned to them. One sluggishly tried to get to its feet.

Shotguns boomed and M3s chattered. The submachine gun sent tremors up Erhardt's arms as he sprayed the beasts. Bloody holes spurted across their bodies. One had the left side of its head explode in a cloud of red.

The guns fell silent. All four Mothmen lay dead.

"Well . . ." Heldmann lowered his shotgun. "That was much easier than the last time we fought them."

Erhardt nodded and took a couple of steps toward the portal. He just stared at it. What lay on the other side? More Mothmen, for sure. But how many? What was their world like? Their society?

Just how smart were they? Did they possess technology more advanced than humans? Or maybe they had not progressed beyond, say, a Bronze Age level of civilization, and just stumbled across some naturally occurring portals.

For fifty years these questions haunted him. How many times did he wish *Rittmeister* von Richthofen had set aside caution and allowed them to enter the portal? The intelligence they might have gained could have prevented the disaster at the Silver Bridge, saved all those people.

And they will be back again. I know they will. And we will know no more about the Mothmen than we do now.

Unless . . .

"Okay," said Jones. "Let's set the charges and seal up this cave."

"No," muttered Erhardt.

"No?" Jones said in a stunned tone. "What do you mean no? You want more of these things getting loose? Maybe dropping another bridge?"

Erhardt swung around. "What do we know about these creatures? What do we actually know?"

Several seconds of silence passed before he continued, "Exactly. Most of what we know about the Mothmen is speculation. We know nothing of their civilization. We do not know how many of these portals are scattered around the world. We do not know what their future plans for us are. Will they keep sending just a few Mothmen to our world, or will they one day decide to launch a full-scale invasion? Do they have weapons on par with ours, or even better?"

His face stiffened. "We need to know, and there is only one way to do that." He looked over his shoulder at the portal.

Heldmann's eyes widened. "You cannot be serious, Fritz. There could be a hundred Moth--"

"Yes," Erhardt cut him off. "There could be a hundred or a thousand Mothmen on the other side of that portal. That concern is what prevented us from going through it fifty years ago. We cannot pass up this opportunity."

Jones let out a breath. "Well, if that's the case, I've got some 'Lurps' in the Shadow Hunters who'd be perfect for this mission." He used the slang for the U.S. Army's Long Range Reconnaissance Patrol.

"No. Better to risk one man than several."

"Then let me go, sir." Kirkland stepped up to him. "I have more experience and training in ground operations than you do. I'm also more expendable."

Erhardt grinned. "Thank you, Master Sergeant, but this is something I have to do." His jaw tightened for a moment. "Something I should have done fifty years ago."

"You mean something *we* should have done." Heldmann rested the shotgun on his shoulder. "I was in the trenches with you that day, too. Besides, *Rittmeister* von Richthofen did not want us to fight alone. You need a wingman."

"Not this time, *mein Freund.*" He clasped Heldmann's shoulder. "There are so few of us left from the Flying Circus. I need someone who has practical experience fighting the Mothmen to stay behind, and that must be you."

Heldmann's shoulders sagged and he stared at the cave floor. Several seconds passed before he raised his head and gazed into Erhardt's eyes. "Then you had better come back. I'll be damned if I have to explain to your son why I did not stop you from doing something foolish."

"Danke." Erhardt smiled.

"You should take these, sir." Weber handed him two extra magazines for his M3. "You might need them."

Heldmann handed over a pair of M1911 magazines to Erhardt. Kirkland gave him a couple of grenades. Erhardt thanked them all, then turned to General Jones.

"If I am not back in an hour, blow the cave."

Jones' right cheek twitched. "Then you better get your ass back here in fifty-nine minutes, fifty-nine seconds, ya dumb Kraut."

Erhardt softly chuckled, then turned to the portal. His stomach clenched as he stared at the glowing blue circle, his imagination stirring up all sorts of possible images of the Mothmen's home world. He did not think it would look like Hell. Unlike when the Mothmen first appeared, he no longer believed they were actual demons, like the one that had possessed his sister Claudia. He had had other experiences in the *Schatten Jäger* that convinced him of that.

He clutched the submachine gun tighter, thought about the camera in his backpack. He wondered what his photo analysts would make of the pictures he brought back.

If I come back.

Drawing a deep breath, Erhardt stepped into the unknown.

AUTHOR'S NOTES

History's greatest fighter pilot . . . The world's most terrifying cryptid . . . in a fight to the death!

How the hell did I come up with that?

Well, having historical figures face the unknown is not something new. One of the best examples is the book, and later movie, *Abraham Lincoln: Vampire Hunter* by Seth Grahame-Smith. There is also the celebrated Jane Austen novel *Pride and Prejudice* that Grahame-Smith turned into the novel *Pride, Prejudice, and Zombies* that saw protagonist Elizabeth Bennet go from a young woman struggling with love and her own prejudices to a kick-ass zombie fighter.

But the spark of inspiration for this book came from scrolling through Facebook one day and coming upon an ad for the movie *Helen Keller vs. Nightwolves.* I obviously could not resist watching the trailer, which was one of the most bat-shit crazy movie previews I've ever seen. Laughing my ass off, I said half-seriously, "One day, I'm gonna write a book about some historical figure fighting some supernatural threat."

But that off-handed comment stuck with me and festered, and soon developed into a serious idea. But what historical figure to use? And what strange creature should he or she fight?

As it turned out, the idea came during a time when my interest in World War I was growing. A World War II buff for most of my life – to the point I was inspired to write the non-fiction book *Weird and Interesting Stuff from World War II* – around 2016 I stumbled across the TimeGhost YouTube channel when it was in the middle of its *Great War Week-by-Week* series hosted by historian Indy Neidell. Along with being informative, Neidell's personality, passion, and exuberance on the subject matter instantly drew me in.

A year after that series wrapped, one of my favorite metal bands, Sabaton, released their album *The Great War*, with every song dealing with some aspect of World War I. This included Lawrence of Arabia ("Seven Pillars of Wisdom"), Sergeant Alvin York ("82nd All the Way"), and . . . "The Red Baron."

If this story was going to be set during the First World War, the go-to historical figure would have to be the Red Baron. Even people with only scant knowledge about that conflict know the name Red Baron. While they may not know the life story of Manfred von Richthofen, they know his nickname is synonymous with a great fighter pilot.

So what monster should I have the Red Baron fight? As a pilot, it would have to be a creature that can fly. As a major cryptozoology enthusiast, I thought about the flying cryptids reported throughout the decades and kept coming back to one that I felt would be a worthy adversary for the Red Baron.

The Mothman.

Jet black, soaring through the air, and piercing the darkness with those blazing red eyes, the Mothman terrorized the residents of Point Pleasant, West Virginia from November 1966 to December 1967. The first sighting was by two couples driving past the TNT Area. After that night, more than one hundred people claimed to have seen the creature. Local newspapers began publishing stories of the sightings, which were soon picked up by national news outlets.

Following the collapse of the Silver Bridge on December 15th, 1967, no more Mothman sightings were reported in Point Pleasant. Years later, radio journalist and UFOlogist John Keel released the book *The Mothman Prophecies,* which returned the cryptid to the national spotlight. It also attempted to connect the Silver Bridge collapse to the Mothman.

So did some strange flying monster make Point Pleasant its home during 1967-68 and bring down a bridge? Skeptics point out many of the sightings could be people misidentifying a known animal such as a barn owl or a sandhill crane, which can be as tall as an average man and does have red around its eyes.

As for the Silver Bridge, examination of the wreckage determined fractured eyebar set off a chain reaction that caused the bridge to collapse, resulting in 46 deaths.

But the legend of the Mothman did not end there. Decades after its first appearance in West Virginia, sightings of similar creatures have been reported in Chicago, Minneapolis – prior to the collapse of the I-35 Bridge in 2007 – New York City before the terrorist attacks of September 11th, 2001, and Chernobyl immediately after the 1986 reactor explosion. The YouTube channel Donovan Dread has also featured stories from people in West Virginia who have allegedly seen Mothman-like creatures in that state.

As some believers in Mothman theorize that it is either attracted to or causes great tragedy, such a creature would be at home over the bloody battlefields of World War I. And one of the bloodiest battles of that war was fought on the fields of Passchendaele.

While it does not get the same notoriety as the Battles of Verdun or the Somme, Passchendaele – also called the Third Battle of Ypres – came to symbolize the slaughter and waste of life associated with the Great War.

Beginning on July 31st, 1917 with the attack on Pilckem Ridge, the battle lasted until November 10th. The exact number of casualties during the three-plus months of fighting is not known, but estimates put the number between 500,000 and 800,000 combined for both sides. Thousands upon thousands of soldiers did not meet their end by bullet or shell or gas, but by sinking into muddy fields and drowning. By the end of the battle, the Allies had gained six miles of land, which was given up just a few months later in the face of a German offensive.

Passchendaele is a battle that piqued my interest long before Indy Neidell came out with his *Great War Week-by-Week* series. In 2003, the band Iron Maiden released the album *Dance of Death* which included the song "Passchendaele." The brilliant lyrics perfectly described the horror and tragedy of that battle, told through the eyes of a soldier who accepted the fact he would not survive. Though many Maiden fans would declare "Aces High" or "The Trooper" as their favorite song, to me it is "Passchendaele." And as the Red Baron and his Flying Circus were stationed in the area during that battle, I could incorporate the subject of my favorite Iron Maiden song into this book.

Along with Manfred von Richthofen, I tried to include many actual members of his Flying Circus in this book. While the Red Baron is well known, most of his fliers are not. But given his success in battle, von Richthofen was allowed to pick the cream of the crop of the German Air Service to fill the ranks of his unit. As such, many of his men were accomplished fighter pilots in their own right.

I have to start with Werner Voss. When researching him, all I could think was, "This guy was the World War I version of

Maverick from *Top Gun.*" As seen in this book, Voss had little regard for authority, leadership, and discipline. But he was a highly skilled pilot who achieved 46 kills during the war.

Voss, however, was shot down and killed on September 23rd, 1917. Flying in a Fokker triplane, he took on eight British fighters by himself. He hit every enemy plane with machine gun fire, forcing some to retreat from the battle, before Lt. Arthur Rhys Davids delivered the fatal blow to Voss. After the battle, the British flier called him, "The bravest German airman." Voss was 20 years old at the time of his death.

Alois Heldmann survived the war, finishing his service with 15 kills. He passed away on November 1st, 1983 at the age of 87.

Kurt-Bertram von Döring also survived the First World War, ending up with eleven kills. He spent the 1920s serving as an advisor for the Argentine and Peruvian air forces before returning to Germany in the 1930s to join the *Luftwaffe.* He rose to the rank of lieutenant general before the war's end and passed away on July 9th, 1960.

Hans Adam, the "old man" of the unit at age 31, was shot down and killed on November 15th, 1917, leaving behind a wife and two children. He had 21 kills to his credit.

Eduard Dostler also died in battle on August 21st, 1917 when his plane exploded in mid-air after being hit by gunfire from a British fighter. At the time of his death, Dostler had 26 kills.

General Friedrich Sixt von Armin continued to lead the 4th Army after the Battle of Passchendaele and took part in the Spring Offensive of 1918, Germany's final attempt to defeat the Allies . . . or the Entente, if you prefer. He left the military in 1919 following the Army's demobilization and spent the next several years as a public speaker. Von Armin passed away on September 30th, 1936 at age 84.

Finally, we come to the main character of this book, the Red Baron, or as the Germans called him during the war *Der Rote Kampfflieger* – "The Red Fighter Pilot." The nickname Red Baron was created by the Allies as his Prussian title of *Freiherr,* or Free Lord, loosely translates to baron in English. Also, while von Richthofen is associated with the iconic red Fokker triplane, at the time this book takes place, July-August 1917, he was flying the Albatros D.V. Von Richthofen did not start flying the triplane regularly until September 1917.

He achieved 80 kills before his death in combat on April 21st, 1918, roughly two weeks before his 26th birthday. The circumstances of his death have been hotly debated for years. Originally, Canadian pilot Captain Arthur "Roy" Brown was credited with downing the Red Baron. But more research revealed that an Australian Sergeant named Cedric Popkin was manning a machine gun on the ground as von Richthofen was flying over and fired the fatal shot into his chest. Though a scourge to Allied airmen, when the Red Baron's body was recovered from the crash site, the Australian Flying Corps buried him with full military honors, a testament to the respect he had gained among his enemies.

Von Richthofen's death has also been studied by neurologists over the past several years. Given the change in his behavior during his final months, and the headaches and nausea that plagued him after returning from every mission, it is believed the head injury he suffered in the summer of 1917 resulted in a traumatic brain injury. Researchers cited how von Richthofen, who stressed taking no unnecessary risks in battle, violated his own tenets of battle more and more following his head injury. One of which was not flying low to avoid enemy ground fire. But that is what he did on that fateful day in April when he apparently met his demise at the hands of an Aussie machine gunner.

More than 100 years have passed since the death of Manfred von Richthofen, yet his legacy remains stronger than ever. Whether it be in songs, books, movies, or even pizza, the name of the Red Baron is certain to endure for 100 more years.

As for his adversary in this book, for all we know, the Mothman may still be lurking somewhere out there.

About the Author: John J. Rust is a New Jersey native who graduated from Mercer County Community College and the College of Mt. St. Vincent with degrees in broadcasting and communications. After working for New Jersey 101.5 FM, he moved to Arizona and became the sports director at KYCA radio and does play-by-play for high school and college sports. Rust has authored 22 books and several short stories and freelance articles. You can follow John J. Rust at www.facebook.com/johnjrustauthor, Twitter @JohnJRust, and Instagram at johnrust70.

Other Books by John J. Rust

Jack Rastun cryptid-hunting series
Sea Raptor
Reptilian
Demon Flyer
Burrunjor

Fallen Eagle trilogy
Fallen Eagle: Alaska Front
Fallen Eagle: Liberty's Blood
Fallen Eagle: Scorched Earth

Operation Beast Slayer
War of the Worlds: Retaliation (with Mark Gardner)
Dark Wings
Weird and Interesting Stuff from World War II

The Best Phillies Team Ever
The Best Red Sox Team Ever
The Best Dodgers Team Ever
Arizona's All-Time Baseball Team
New Jersey's All-Time Baseball Team
Pennsylvania's All-Time Baseball Team
Here and Gone: Short-Lived Sports Teams and Leagues
The Hindsight Draft: Who MLB Teams Could Have Picked
You're Called the What? Weird, Funny, and Dumb Team Names

RED PORTAL

A PORTAL TO THE PAST . . .

COULD GIVE NORTH KOREA ULTIMATE VICTORY

A mysterious portal suddenly appears along the DMZ, one that leads to the early days of the Korean War. When North Korea discovers it, an armored regiment is sent to the past with one objective. Destroy General MacArthur's invasion force at Incheon and ensure communist rule of the peninsula.

But the U.S. and South Korean militaries learn of the plan and dispatch their own armor and infantry units to 1950. War machines of past and present collide as tank commander Gabe Elarton, military history professor Alicia Vargas, and Black Berets Park and Kwang wage a desperate battle to prevent the North Koreans from changing the course of history.

But it is not only the fate of the peninsula at stake. Should Incheon fall, it could ignite World War III.

OPERATION BEAST SLAYER

While on a NATO exercise in Norway, ace Marine chopper pilot Makato McShane discovers a threat unlike any humanity has ever faced . . . a race of stone giants. Thirty feet tall, impervious to missiles and shells, they lay waste to one city after another.

At the height of the conflict, another monster appears, one straight out of legend. A fire-breathing dragon! But is it friend or foe to humanity?

McShane and mythology professor Ylva Tande team up to find the origin of these creatures and hopefully a way to stop them. Should they fail, Norway, and all of Scandinavia, could face utter devastation.

DEMON FLYER

For decades a terrifying flying monster has been reported throughout Indonesia. It is known as the Ropen, the local term for "Demon Flyer." A survivor from prehistoric times, it is large, deadly . . . and hungry for human flesh.

When fishing boats go missing in the Java Sea, Jack Rastun, Karen Thatcher, and their team of monster hunters from the FUBI are brought in to track down the beast. They soon learn there is not one Ropen, but an entire colony. But standing in their way is a group of activists determined to preserve the Ropen by any means necessary.

With politics and social media outrage tying his hands, Rastun must find a way to overcome his human adversaries or watch the Demon Flyers devour the population of an entire island . . . along with him and his team.

Praise for John J. Rust's Books

"I really enjoyed this book. His character development and multi-layered plot held my interest. There were some superb action sequences that spanned multiple chapters and kept the action flowing, and I genuinely cared about the characters in jeopardy." – *Steve Yeager, author of "Raptor Apocalypse," on "Sea Raptor."*

"If you want a monster-chasing adventure, this is an exciting book." – *Matt Bille, author of "The Dolmen" on "Sea Raptor."*

"A must read for action/horror fans." – *Amazon review for "Reptilian."*

"This was Rust's best FUBI novel yet." -- *Amazon review for "Demon Flyer."*

"If military fiction has a place in your reading list, read this one." -- *Andy McKinney of the Payson Roundup on "Fallen Eagle: Alaska Front."*

"John Rust creates an excellent combination of thoughtful speculation and excellent action. This is an exciting read that also leaves you thinking." – *Amazon Review for "Operation Beast Slayer."*

"The action never stops, neither does the dino." – *Amazon Review for "Burrunjor"*

www.ingramcontent.com/pod-product-compliance
Lightning Source LLC
Chambersburg PA
CBHW061219170626
46809CB00007B/2525